INTER/FACE

HAZARD
CLASSIFICATION
SYSTEMS

A Comparative Guide to Definitions and Labels

HAZARDOUS MATERIALS

INFORMATION CENTER

A Division of Inter/Face Associates, Inc.
62 Washington Street
Middletown, Connecticut 06457

First printing 1986

Although Inter/Face Associates has conducted extensive research to ensure the accuracy and completeness of the information contained in this book, we assume no responsibility for errors, inaccuracies, omissions or any inconsistency herein.

The views and opinions expressed in this book are solely those of the publisher. The National Fire Protection Association was not involved in the preparation of nor has it reviewed or approved the text of this book and, therefore, takes no responsibility for the views, instructions, opinions or conclusions expressed herein.

Certain identified material in Chapter 4 hereof is reprinted with permission from NFPA 704 - 1985, Standard System for the Identification of the Fire Hazards of Materials, Copyright ©1985, National Fire Protection Association, Quincy, Massachusetts 02269. This reprinted material is not the complete and official position of the NFPA on the referenced subject which is represented only by the standard in its entirety.

The NFPA hazard classification warning system is intended to be interpreted and applied only by properly trained individuals to identify fire, health and reactivity hazards of chemicals. The user is referred to a certain limited number of chemicals with recommended classifications in NFPA 49 and NFPA 325M which should be used as a guideline only. Whether the chemicals are classified by NFPA or not, anyone using the 704 system to classify chemicals does so at their own risk.

Library of Congress Cataloging-in-Publication Data

Hazard Classification Systems.

    1. Hazardous substances-—United States—Classification.
2. Hazardous substances—Labeling—United States.
3. Hazardous substances—Law and legislation—United States.
I. Inter/Face Associates.
T55.3.H3H387 1986   363.1 ' 762 ' 0973   86-15199
ISBN 0-938135-04-X

Printed in the United States of America

CONTENTS

CHAPTER 3 – EPA HAZARDOUS WASTES

Contents of Chapter
EPA Hazardous Wastes

EPA's Definitions Compared to DOT, OSHA and NFPA

CHAPTER 4 – TEXTS OF HAZARD DEFINITIONS

Contents of Chapter

APPENDIX A – DEPARTMENT OF TRANSPORTATION REGULATIONS ON MULTIPLE LABELING AND THE HAZARD CLASS HIERARCHY

This Guide is designed to assist the regulated community in determining whether or not a chemical or material is regulated under any of three federal programs. With this Guide, a label or the results of an analysis from one federal program can be used to identify chemicals covered under another program.

The Guide analyzes and compares the hazard definitions used by three major federal programs. The programs covered in the Guide are:

- ☐ the Department of Transportation (DOT) program for regulating the transportation of hazardous materials;

- ☐ the Occupational Safety and Health Administration (OSHA) program for regulating employee access to information about hazardous chemicals in the workplace (the Hazard Communication Standard); and

- ☐ the Environmental Protection Agency (EPA) program for regulating the management of hazardous waste.

In addition, the Guide compares the hazard definitions used by these programs to the definitions used by the National Fire Protection Association (NFPA). The NFPA uses its definitions to assign values used in its "diamond" symbol which provides a quick indication of the fire hazards of materials.

Thousands of chemicals and other materials are regulated as hazardous by the federal government. Many of these materials are subject to regulation by more than one federal agency. Identifying these materials is relatively simple for the ones that have been specifically listed by an agency.

The majority of regulated chemicals, however, are not listed. The task of identifying these un-listed chemicals is more complicated, since it is based on the definitions of hazard classes used by the various regulatory agencies. Each federal agency establishes definitions of various hazards under its jurisdiction and regulates all chemicals which meet the criteria of any of the definitions. It is up to the regulated community to determine whether or not a particular chemical meets any of the definitions and is, therefore, regulated. What makes this particularly difficult is that there is little uniformity among the federal agencies in the definitions they use. Each agency evaluates hazards from its own perspective, based on its own regulatory responsibilities. As those responsibilities differ from agency to agency, so do the hazard definitions they develop. In many cases, however, close analysis reveals that these differences are more in form than in substance.

For each of the three federal programs and for each hazard class defined by the relevant agency, the Guide provides a summary chart and a narrative analysis. The Guide points out the similarities and differences among the hazard classification systems for all of the commonly used hazard classes.

Continued...

The summary chart is a quick reference to how a hazard class in one program compares to a hazard class in another. Illustrations of the label required by DOT for hazardous materials and the NFPA diamond for the hazard class are included to facilitate comparison and identification.

The narrative analysis includes a more detailed description of the hazard class and an analysis of how it compares to corresponding hazard classes in the other programs.

Chapter 1 compares the hazard classes developed by OSHA for the Hazard Communication Standard to corresponding hazard classes developed by DOT, EPA and NFPA. Chapter 2 compares DOT's hazard classes to those of OSHA, EPA and NFPA. Chapter 3 compares EPA's characteristic waste hazard classes to DOT, OSHA and NFPA classes. Each chapter is designed so that information available to the regulated community under the subject agency's program can be used to determine if the material is covered by the other agencies' programs and to identify its hazard class or classes under those programs.

The hazard class comparisons should only be used for unlisted substances. If a material is listed by DOT in the Hazardous Materials Table (49 CFR 172.101), the proper hazard class and required label(s) are identified in the table. If a waste is listed by EPA in the "F" list (40 CFR 261.31), the "K" list (40 CFR 261.32), the "P" list (40 CFR 261.33) or the "U" list (40 CFR 261.33), it is a hazardous waste with the characteristic and hazardous waste number assigned in the list. If a material is listed in *NFPA 49: Hazardous Chemical Data* or *NFPA 325M: Fire Hazard Properties of Flammable Liquids, Gases, and Volatile Solids*, the fire, reactivity, health and special hazard classes assigned in the lists should be used. OSHA specifies two lists which establish that a chemical is a hazardous chemical, but it still requires the chemical manufacturer to evaluate the chemical to determine its chemical and physical properties and physical and health hazards. Therefore, the hazard class comparisons in this Guide are appropriate for all OSHA hazardous chemicals.

It is important to note that, of the four hazard classification systems analyzed, only the DOT has a ranking system for determining the appropriate hazard class for a material that meets the criteria of more than one hazard class. This ranking system is used to determine the proper shipping requirements for a material and the proper shipping label(s) for unlisted, multiple hazard class materials. For many multiple hazard class combinations, DOT requires multiple labels. The DOT hazard class ranking system and the regulations covering multiple labeling are presented in the Appendix to this Guide.

The final chapter of the Guide is the actual text of the definitions for various hazard classes from the various agencies and the NFPA. The texts are current as of 1 June 1986.

* * *

OSHA HAZARDOUS CHEMICALS

CONTENTS OF CHAPTER

OSHA HAZARDOUS CHEMICALS

This chapter compares the hazard classification system developed by the Occupational Safety and Health Administration (OSHA) for the Hazard Communication Standard (29 CFR 1910.1200) to the systems developed by the Department of Transportation (DOT) for the transportation of hazardous materials, the Environmental Protection Agency (EPA) for the management of hazardous waste and the National Fire Protection Association (NFPA) to identify the fire hazards of materials.

OSHA's Hazard Communication Standard, informally known as employee right-to-know regulations, requires employers to provide information on chemical hazards in the workplace to workers and to their doctors, to customers who purchase hazardous chemicals and to certain other parties. The Standard went into effect for chemical manufacturers, distributors, and importers in November 1985 and for all other manufacturers in May 1986.

Under the Standard, chemical manufacturers (manufacturers who produce a regulated chemical for use or distribution) are required to evaluate the chemicals they produce to determine which of them are hazardous. OSHA's definition of "hazardous chemical" includes chemicals that are listed in any of four reference lists as well as chemicals meeting the criteria of any of OSHA's 18 physical and health hazard definitions.

Chemical manufacturers must prepare or obtain Material Safety Data Sheets (MSDSs) for the chemicals they determine to be hazardous and provide the MSDSs to those who purchase their chemicals. An MSDS must include a description of the physical and chemical characteristics of the chemical as well as its physical and health hazards.

Because an MSDS must contain this information about hazardous chemicals and because one must be provided for every hazardous chemical, MSDSs are an excellent source of information that can be used to simplify compliance with the DOT and EPA programs and use of the NFPA labeling system. This chapter is designed to assist the regulated community in using the information on an MSDS for these purposes.

For each of the hazard classes defined by OSHA in the Hazard Communication Standard, the chapter identifies the corresponding hazard class defined by DOT, EPA and NFPA. This chapter also identifies the DOT label appropriate for the hazard class and the NFPA fire, reactivity, health and special hazard rating for the class.

The hazard class comparisons should only be used for unlisted substances. If a material is listed by DOT in the Hazardous Materials Table (49 CFR 172.101), the proper hazard class and required label(s) are identified in the table. If a waste is listed by EPA in the "F" list (40 CFR 261.31), the "K" list (40 CFR 261.32), the "P" list (40 CFR 261.33) or the "U" list (40 CFR 261.33), it is a hazardous waste with the characteristic and hazardous waste number assigned in the list. If a material is listed in *NFPA 49: Hazardous Chemical Data* or *NFPA 325M: Fire Hazard Properties of Flammable Liquids, Gases, and Volatile Solids*, the fire, reactivity, health and special hazard classes assigned in the lists should

Continued . . .

be used. OSHA specifies two lists which establish that a chemical is a hazardous chemical covered by the Standard, but OSHA still requires the chemical manufacturer to evaluate the chemical to determine its physical and chemical properties and its physical and health hazards. This caution about use of the comparisons, therefore, does not apply to OSHA hazardous chemicals.

It is important to note that, of the four hazard classification systems analyzed, only the DOT has a ranking system for determining the appropriate hazard class for a material that meets the criteria of more than one hazard class. This ranking system is used to determine the shipping requirements for a material and the proper shipping label(s) for unlisted, multiple hazard class materials. For many multiple hazard class combinations, DOT requires multiple labels. The DOT hazard class ranking system and the regulations covering multiple labeling are presented in the Appendix to this Guide. To use the OSHA/DOT comparisons in this chapter, identify all of the DOT hazard classes that correspond to the physical and health hazards on the OSHA MSDS. If only one DOT hazard class applies, use the hazard class and label indicated in the analysis of the hazard class contained in this chapter. If more than one DOT hazard class applies, consult the Appendix to determine the appropriate hazard class for the material and the label(s) to be used.

Hazardous wastes regulated by EPA are specifically exempted from regulation under the OSHA Hazard Communication Standard. However, the information about a hazardous chemical provided on an MSDS can be used to determine whether or not the chemical is a hazardous waste if it becomes a waste. The analysis in this chapter is designed to assist the regulated community in using MSDS information about a chemical that becomes a waste to determine if it is a hazardous waste and the characteristic(s) that make it a hazardous waste. Wastes that are determined to be hazardous wastes because they have one or more of the hazardous characteristics defined by EPA are assigned hazardous waste numbers based on the characteristic. The hazardous waste number for ignitable wastes (gases, liquids, solids and oxidizers) is D001; for corrosive wastes, D002; and for reactive wastes (cyanide and sulfide bearing wastes, explosives, normally unstable wastes and water reactives) is D003. EPA does not have rules about which characteristic to use if a waste meets the criteria of more than one characteristic.

The NFPA diamond is designed to convey information about all of the hazards of a material. The top portion of the diamond is for flammability hazards, the right portion is for reactivity hazards, the left portion is for health hazards and the bottom portion is for special hazards. The degree of hazard is indicated by numbers ranging from 0 to 4, with 4 denoting the highest degree of hazard. The analysis in this chapter identifies the NFPA hazard class and degree(s) that correspond to each OSHA hazard class. For chemicals that have more than one hazard, the NFPA class and degree should be determined for each hazard. The proper label for the material consists of the appropriate degree of hazard in each portion of the diamond. If more than one degree is identified for a particular NFPA hazard class, use the highest number.

* * *

4

OSHA'S DEFINITION OF "CARCINOGEN"
COMPARED TO DOT, EPA AND NFPA DEFINITIONS

OSHA		DOT
"CARCINOGEN"	=	NO CORRESPONDING DEFINITION

OSHA		EPA
"CARCINOGEN"	= *PROBABLE*	"TOXIC WASTE"

Most OSHA carcinogens are EPA toxic wastes,
but many EPA toxic wastes are not OSHA carcinogens.

OSHA		NFPA
"CARCINOGEN"	=	NO CORRESPONDING DEFINITION

OSHA'S DEFINITION OF "CARCINOGEN"
COMPARED TO DOT, EPA AND NFPA DEFINITIONS

OSHA defines a chemical as a "carcinogen" if it is regulated as a carcinogen by OSHA or if it is listed as a carcinogen or potential carcinogen by either the National Toxicology Program (NTP) in its *Annual Report on Carcinogens* or by the International Agency for Research on Cancer (IARC) in any of its *Monographs on the Evaluation of the Carcinogenic Risk of Chemicals to Humans.*

DOT does not have a definition of a hazard that corresponds to OSHA's definition of "carcinogen."

EPA does not define "carcinogen" in its hazardous waste management regulations but it does list carcinogenicity as one of the criteria to be used in listing a substance in "Appendix VIII" — Hazardous Constituents" to "Part 261 — Identification and Listing of Hazardous Waste" of its regulations. Substances are to be listed in Appendix VIII if scientific studies have shown them to be, among other things, carcinogens. If a substance listed in Appendix VIII is present in a waste, the waste must be listed by EPA as a hazardous waste unless EPA determines that the waste does not pose a hazard to human health or the environment if improperly managed. Wastes listed because of the presence of a substance listed in Appendix VIII are "toxic wastes." Because EPA does not specify particular sources for "scientific studies," some OSHA carcinogens may not be listed by EPA in Appendix VIII and, conversely, some substances listed by EPA in Appendix VIII because of carcinogenicity may not be OSHA carcinogens. In addition, some substances listed in Appendix VIII because of carcinogenicity may not be listed as toxic wastes because EPA has determined that they will not pose a hazard if improperly managed. Most substances that are OSHA carcinogens are listed EPA toxic wastes if they become waste but there is not a one-to-one correspondence between the two terms.

NFPA does not define a hazard that corresponds to OSHA's definition of "carcinogen."

OSHA'S DEFINITION OF "COMBUSTIBLE LIQUID"
COMPARED TO DOT, EPA AND NFPA DEFINITIONS

OSHA   **DOT**

"COMBUSTIBLE LIQUID"   =

All OSHA combustible liquids are DOT combustible liquids ("Combustible" placard)
and all DOT combustible liquids are OSHA combustible liquids.

OSHA   **EPA**

"COMBUSTIBLE LIQUID"   =   **"IGNITABLE LIQUID"**

*FOR FLASH POINTS
BELOW 140°F ONLY*

OSHA combustible liquids with a flash point below 140°F are EPA ignitable liquids;
EPA ignitable liquids with a flash point between 100°F and 140°F are OSHA combustible liquids.

OSHA   **NFPA**

"COMBUSTIBLE LIQUID"   =

All OSHA combustible liquids are NFPA Flammability Hazard 2 materials,
but only NFPA Flammability Hazard 2 liquids are OSHA combustible liquids.

OSHA'S DEFINITION OF "COMBUSTIBLE LIQUID" COMPARED TO DOT, EPA AND NFPA DEFINITIONS

OSHA's definition of "combustible liquid" focuses on the flash point of the liquid. Liquids with a flash point at or above 100°F and below 200°F are combustible liquids.

DOT's definition of "combustible liquid" is essentially identical to OSHA's defintion of "combustible liquid." Any liquid that is a combustible liquid under OSHA's Hazard Communication Standard is a DOT "combustible liquid" if the liquid does not qualify for a higher DOT hazard class. All DOT combustible liquids are OSHA combustible liquids. DOT does not require a label on packages containing a combustible liquid. However, when combustible liquids are transported in packages with a capacity of greater than 110 gallons or in a cargo tank or tank car, a "combustible" placard must be used.

EPA's definition of "ignitable liquid" also focuses on the flash point of the liquid, but it covers liquids with a flash point less than 140°F. OSHA's definition of "combustible liquid" covers liquids with a flash point between 100°F and 200°F. Therefore, only OSHA combustible liquids with a flash point less than 140°F are EPA ignitable liquids if they become waste. Conversely, only EPA ignitable liquids with a flash point between 100°F and 140°F are OSHA combustible liquids before they become waste.

NFPA's Flammability Hazard 2 includes liquids with a flash point range that is the same as in OSHA's definition of "combustible liquid." Flammability Hazard 2 also includes solids and semi-solids which readily give off flammable vapors. Therefore, all OSHA combustible liquids are NFPA Flammability Hazard 2 materials but only liquids with an NFPA Flammability Hazard 2 are OSHA combustible liquids.

OSHA'S DEFINITION OF "COMPRESSED GAS"
COMPARED TO DOT, EPA AND NFPA DEFINITIONS

OSHA   DOT

"COMPRESSED GAS"   =
NON-FLAMMABLE GAS

All OSHA compressed gases are DOT compressed gases ("Non-Flammable Gas" label),
and all DOT compressed gases ("Non-Flammable Gas" label) are OSHA compressed gases.

OSHA   EPA

"COMPRESSED GAS"   =   NO CORRESPONDING
DEFINITION

OSHA   NFPA

"COMPRESSED GAS"   =   NO CORRESPONDING
DEFINITION

OSHA'S DEFINITION OF "COMPRESSED GAS"
COMPARED TO DOT, EPA AND NFPA DEFINITIONS

OSHA's definition of "compressed gas" focuses on the pressure of a gas in a container at various temperatures.

DOT's definition of "compressed gas" is essentially the same as OSHA's definition. Compressed gases that are not flammable must be labeled with the DOT "Non-Flammable Gas" label. All OSHA compressed gases are DOT compressed gases and vice versa.

Neither **EPA** nor **NFPA** have definitions of hazards that correspond to OSHA's definiton of "compressed gas."

OSHA'S DEFINITION OF "CORROSIVE"
COMPARED TO DOT, EPA AND NFPA DEFINITIONS

OSHA   **DOT**

"CORROSIVE"   =

All OSHA corrosives are DOT corrosive materials,
but DOT "metal" corrosive materials are not OSHA corrosives.

OSHA   **EPA**

"CORROSIVE"   =   **"CORROSIVITY"**

PROBABLE

Most OSHA corrosives are EPA "pH" corrosives,
but EPA "metal" corrosives are not OSHA corrosives.

OSHA   **NFPA**

"CORROSIVE"   =

All OSHA corrosives are NFPA Health Hazard 3 materials,
but not all NFPA Health Hazard 3 materials (toxics and toxic combustion products) are OSHA corrosives.

OSHA'S DEFINITION OF "CORROSIVE"
COMPARED TO DOT, EPA AND NFPA DEFINITIONS

OSHA's definition of "corrosive" focuses on the destructive effects of a substance on human tissue. The test specified to determine if a chemical is corrosive involves applying the chemical to the skin of an albino rabbit.

DOT's definition of "corrosive material" includes a "skin" component and a "metal" component. The "metal" component concerns the corrosion rate of a substance on metal. The test specified to determine if a chemical is corrosive to metal involves applying the chemical to steel. The "skin" component of DOT's definition of "corrosive material" is essentially identical to OSHA's definition of "corrosive." Substances that have a DOT hazard class of "corrosive material" because of the "skin" portion of the definition are corrosive for the Hazard Communication Standard. Containers of substances in the corrosive material hazard class must have the DOT "corrosive" label. All OSHA corrosives are DOT corrosive materials unless the material also qualifies for a higher hazard class. Not all DOT corrosive materials (i.e., the "metal" corrosives) are OSHA corrosives.

EPA's definition of "corrosivity" includes a "metal" component identical to DOT's and a "pH" component. The "pH" component of EPA's definition specifies that wastes with a pH of 2 or less, or a pH of 12.5 or more are corrosive hazardous wastes. Although OSHA does not have a definition that corresponds to the "pH" component of EPA's definition, most chemicals that have a pH of 2 or less, or a pH of 12.5 or more will be classed as corrosive by the "skin" test specified in OSHA's definition of "corrosive." However, there may be chemicals with a pH of between 2 and 12.5 that are corrosive by the "skin" test, just as there may be chemcials with a pH less than 2 or greater than 12.5 that would not be classed as corrosive by the "skin" test. Therefore, while most chemicals that are OSHA corrosives are also corrosive under the "pH" portion of EPA's corrosivity definition, there is not necessarily a one-to-one correspondence.

NFPA's Health Hazard 3 includes materials covered by OSHA's definition of "corrosive." Health Hazard 3, however, also includes materials that are toxic and those that give off highly toxic combustion products. All OSHA corrosives are NFPA Health Hazard 3 corrosives, but not all materials with a Health Hazard 3 are OSHA corrosives.

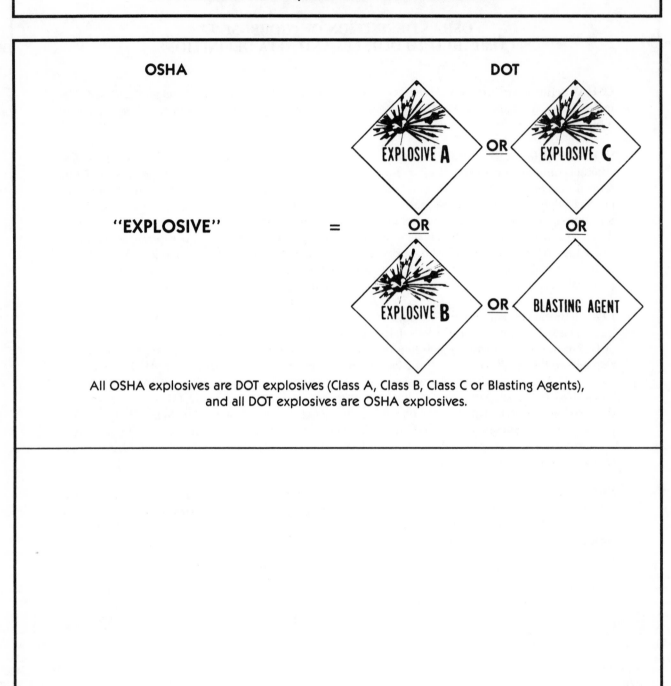

OSHA   **DOT**

"EXPLOSIVE"   **=**

All OSHA explosives are DOT explosives (Class A, Class B, Class C or Blasting Agents), and all DOT explosives are OSHA explosives.

Continued . . .

14

OSHA'S DEFINITION OF "EXPLOSIVE"
COMPARED TO DOT, EPA AND NFPA DEFINITIONS

OSHA's definition of "explosive" is relatively simple. A chemical that causes a sudden, almost instantaneous release of pressure, gas and heat when subjected to a sudden shock, pressure or high temperature is an explosive.

DOT defines three classes of explosives (Classes A, B and C) and a class of blasting agents. Class A explosives are high explosives, detonating explosives and other explosives of maximum hazard. Class B explosives are deflagrating explosives or explosives that react by rapid burning rather than detonating. Class C explosives are manufactured articles that contain restricted quantities of Class A or Class B explosives. Blasting agents are materials used for blasting and have a very low potential for ignition during transportation. Materials in all four of these classes cause a sudden, almost instantaneous release of pressure, gas and heat when subjected to a sudden shock, pressure or high temperature. Therefore, all OSHA explosives are DOT Class A, Class B or Class C explosives or blasting agents. Conversely, all DOT explosives (Class A, B and C explosives and blasting agents) are OSHA explosives under the Hazard Communication Standard. The DOT labels required for the four classes are "Explosive A," "Explosive B," "Explosive C" and "Blasting Agent" respectively.

Continued . . .

OSHA'S DEFINITION OF "EXPLOSIVE"
COMPARED TO DOT, EPA AND NFPA DEFINITIONS

OSHA   EPA

"EXPLOSIVE"   =   "REACTIVE (EXPLOSIVE)"

All OSHA explosives are EPA reactives (explosive)
and all EPA reactives (explosive) are OSHA explosives.

OSHA   NFPA

"EXPLOSIVE"   =   ◇4 OR ◇3

All OSHA explosives are NFPA Reactivity Hazard 4 or Reactivity Hazard 3 materials,
and all NFPA Reactivity Hazard 4 materials are OSHA explosives,
but not all Reactivity Hazard 3 materials (water reactives) are OSHA explosives.

EPA's Reactive (Explosive) characteristic for hazardous waste has three components. First, DOT Class A and Class B explosives are EPA explosives. As indicated in the preceding paragraph, all DOT Class A and B explosives are OSHA explosives. The second component of EPA's reactive (explosive) characteristic is a paraphrase of NFPA's Reactivity 4. A waste is an EPA explosive if it is readily capable of detonation or explosion at standard temperature and pressure. Under the third component, a paraphrase of NFPA's Reactivity 3, a waste is an EPA explosive if it is capable of detonation or explosion when subjected to a strong initiating source or when heated under confinement. The criteria of these two definitions are covered by OSHA's definition of "explosive." All materials that are OSHA explosives meet the criteria for EPA's characteristic of reactive (explosive) when they become waste.

NFPA's Reactivity Hazard 4 covers materials that are readily capable of detonation or explosion at normal temperatures and pressures. Reactivity Hazard 4 includes materials sensitive to mechanical or localized thermal shock at normal temperature and pressure. NFPA's Reactivity Hazard 3 covers materials capable of detonation or explosion when subject to a strong initiating source or when heated under confinement. Reactivity Hazard 3 also includes materials which react explosively with water. The criteria of Reactivity Hazards 4 and 3 correspond to the provisions of OSHA's definition of "explosive." All OSHA explosives are NFPA Reactivity Hazard 4 or Reactivity Hazard 3 materials and all materials with an NFPA Reactivity Hazard 4 are OSHA explosives. Because NFPA Reactivity Hazard 3 includes water reactives in addition to explosives, some materials with a Reactivity Hazard 3 are not OSHA explosives.

OSHA'S DEFINITION OF "FLAMMABLE AEROSOL" COMPARED TO DOT, EPA AND NFPA DEFINITIONS

OSHA   DOT

"FLAMMABLE AEROSOL"   =

All OSHA flammable aerosols are DOT flammable compressed gases ("Flammable Gas" label), but not all DOT flammable compressed gases are OSHA flammable aerosols.

OSHA   EPA

"FLAMMABLE AEROSOL"   =   **"IGNITABLE COMPRESSED GAS"**

All OSHA flammable aerosols are EPA ignitable compressed gases, but not all EPA ignitable compressed gases are OSHA flammable aerosols.

OSHA   NFPA

"FLAMMABLE AEROSOL"   =

All OSHA flammable aerosols are NFPA Flammability Hazard 4 materials, but not all NFPA Flammability Hazard 4 materials are OSHA flammable aerosols.

OSHA'S DEFINITION OF "FLAMMABLE AEROSOL"
COMPARED TO DOT, EPA AND NFPA DEFINITIONS

OSHA defines "flammable aerosol" as an aerosol that, in a Consumer Product Safety Commission test, projects a flame more than 18 inches or has a flame that extends back to the valve.

DOT's definition of "flammable compressed gas" includes a portion essentially identical to OSHA's definition of "flammable aerosol." DOT's definition, however, has three additional components. Therefore, all OSHA flammable aerosols are DOT flammable compressed gases unless the material also qualifies for a higher DOT hazard class. However, not all DOT flammable compressed gases are OSHA flammable aerosols. The DOT label for flammable compressed gases is the "Flammable Gas" label.

EPA's definition of "ignitable compressed gas" simply refers to DOT's definition of "flammable compressed gas." Therefore, the correspondence between OSHA's "flammable aerosol" and EPA's "ignitable compressed gas" is the same as between the OSHA definition and DOT's "flammable compressed gas." If it becomes a waste, an OSHA flammable aerosol is an EPA ignitable compressed gas.

NFPA's Flammability Hazard 4 includes, among other things, flammable gases. Since an OSHA flammable aerosol is a flammable gas, all OSHA flammable aerosols are NFPA Flammability Hazard 4 materials. NFPA Flammability Hazard 4, however, also includes other forms of material. Therefore, not all substances with a Flammability Hazard 4 are OSHA flammable aerosols.

OSHA'S DEFINITION OF "FLAMMABLE GAS"
COMPARED TO DOT, EPA AND NFPA DEFINITIONS

OSHA   **DOT**

"FLAMMABLE GAS"   **=**

All OSHA flammable gases are DOT flammable compressed gases ("Flammable Gas" label),
but not all DOT flammable compressed gases are OSHA flammable gases.

OSHA   **EPA**

"FLAMMABLE GAS"   **=**   **"IGNITABLE COMPRESSED GAS"**

All OSHA flammable gases are EPA ignitable compressed gases,
but not all EPA ignitable compressed gases are OSHA flammable gases.

OSHA   **NFPA**

"FLAMMABLE GAS"   **=**

All OSHA flammable gases are NFPA Flammability Hazard 4 materials,
but not all NFPA Flammability Hazard 4 materials are OSHA flammable gases.

OSHA'S DEFINITION OF "FLAMMABLE GAS" COMPARED TO DOT, EPA AND NFPA DEFINITIONS

OSHA's definition of "flammable gas" has two components. A gas that satisfies the criteria of either of the components is a "flammable gas." Under the first component, a gas is a "flammable gas" if it forms a flammable mixture with air when the concentration of the gas is 13 percent or less. Under the second component, a gas is a "flammable gas" if it has a range of flammable concentrations with air that is wider than 12 percent regardless of the lower limit of its range.

DOT's definition of "flammable compressed gas" includes a provision that is essentially identical to OSHA's definition of "flammable gas." DOT's definition, however, also includes a flammable aerosol provision as well as two others. Therefore, all OSHA flammable gases are DOT flammable compressed gases unless the material also qualifies for a higher DOT hazard class. However, not all DOT flammable compressed gases are OSHA flammable gases. The DOT label for flammable compressed gases is the "Flammable Gas" label.

EPA's definition of "ignitable compressed gas" simply refers to DOT's definition of "flammable compressed gas." Therefore, the correspondence between OSHA's "flammable gas" and EPA's "ignitable compressed gas" is the same as between the OSHA definition and DOT's "flammable compressed gas." If it becomes a waste, an OSHA flammable gas is an EPA ignitable compressed gas.

NFPA's Flammability Hazard 4 includes, among other things, flammable gases. All OSHA flammable gases are NFPA Flammability Hazard 4 materials. NFPA Flammability Hazard 4, however, also includes other forms of material. Therefore, not all substances with Flammability Hazard 4 are OSHA flammable gases.

OSHA'S DEFINITION OF "FLAMMABLE LIQUID" COMPARED TO DOT, EPA AND NFPA DEFINITIONS

OSHA　　　　　　　　　　　　**DOT**

"FLAMMABLE LIQUID"　　　=　　

All OSHA flammable liquids are DOT flammable liquids,
and all DOT flammable liquids are OSHA flammable liquids.

OSHA　　　　　　　　　　　　**EPA**

"FLAMMABLE LIQUID"　　　=　　**"IGNITABLE LIQUID"**

All OSHA flammable liquids are EPA ignitable liquids,
but not all EPA ignitable liquids (flash points of 100°F or greater) are OSHA flammable liquids.

OSHA　　　　　　　　　　　　**NFPA**

"FLAMMABLE LIQUID"　　　=　　

All OSHA flammable liquids are NFPA Flammability Hazard 4 or Flammability Hazard 3 materials,
but not all NFPA Flammability Hazard 4 and 3 materials are OSHA flammable liquids.

OSHA'S DEFINITION OF "FLAMMABLE LIQUID" COMPARED TO DOT, EPA AND NFPA DEFINITIONS

OSHA's definition of "flammable liquid" focuses on the flash point of a liquid. A liquid with a flash point of less than 100°F is an OSHA flammable liquid.

DOT's definition of "flammable liquid" is essentially identical to OSHA's definition. All OSHA flammable liquids are DOT flammable liquids and vice versa. DOT flammable liquids must have the DOT "Flammable Liquid" label.

EPA's definition of "ignitable liquid" includes liquids with flash points of up to 140°F. Therefore, all OSHA flammable liquids are EPA ignitable liquids if they become waste, but not all EPA ignitable liquids (those with a flash point of 100°F or higher) are OSHA flammable liquids.

NFPA's Flammability Hazards 4 and 3 include boiling point criteria in addition to flash point to classify liquids. A liquid with a flash point below 73°F and a boiling point below 100°F is a Flammability Hazard 4. The criteria for Flammability Hazard 3 are 1) a flash point below 73°F and a boiling point at or above 100°F or 2) a flash point at or above 73°F but below 100°F. NFPA Flammability Hazards 4 and 3 also cover gases, solids, dusts and mists. Therefore, while all OSHA flammable liquids are NFPA Flammability Hazard 4 or Flammability Hazard 3 materials, not all materials classed as NFPA Flammability Hazards 4 or 3 are OSHA flammable liquids.

OSHA'S DEFINITION OF "FLAMMABLE SOLID"
COMPARED TO DOT, EPA AND NFPA DEFINITIONS

OSHA

DOT

"FLAMMABLE SOLID" =

All OSHA flammable solids are DOT flammable solids,
but not all DOT flammable solids (water reactives) are OSHA flammable solids.

OSHA

EPA

"FLAMMABLE SOLID" = "IGNITABLE SOLID"

All OSHA flammable solids are EPA ignitable solids,
and all EPA ignitable solids are OSHA flammable solids.

OSHA

NFPA

"FLAMMABLE SOLID" = 4 OR 3 OR 2

All OSHA flammable solids are NFPA Flammability Hazard 4, 3 or 2 materials,
but not all NFPA Flammability Hazard 4, 3 or 2 materials are OSHA flammable solids.

OSHA'S DEFINITION OF "FLAMMABLE SOLID"
COMPARED TO DOT, EPA AND NFPA DEFINITIONS

OSHA's definition of "flammable solid" has two components. The first is that the solid is likely to cause a fire. This can be the result of friction, absorption of moisture, spontaneous chemical change or retained heat from manufacturing. The second component is that the solid can be ignited readily and when ignited burns very vigorously and persistently. OSHA specifies a Consumer Product Safety Commission test for determining if a solid is a flammable solid.

DOT's definition of "flammable solid" is similar to OSHA's but DOT includes water reactive material in its definition while OSHA does not. Therefore, all OSHA flammable solids are DOT flammable solids unless the material qualifies for a higher DOT hazard class. Not all DOT flammable solids (water reactives) are OSHA flammable solids. The DOT label for a flammable solid is the "Flammable Solid" label.

EPA's definition of "ignitable solid" is essentially identical to OSHA's definition of "flammable solid." All OSHA flammable solids are EPA ignitable solids if they become wastes and are, therefore, hazardous wastes. All EPA ignitable solids are OSHA flammable solids before they become wastes.

NFPA does not have a single hazard class that corresponds closely to OSHA's definition of "flammable solid." NFPA Flammability 4 covers materials that will burn readily including dusts of combustible solids that can form explosive mixtures with air. NFPA Flammability Hazard 3 covers solids that burn very rapidly and those that are spontaneously combustible. NFPA Flammability Hazard 2 covers solids that require moderate heating before they ignite but which readily give off flammable vapors. In combination, the criteria for these three classes cover all of the provisions of OSHA's definition of "flammable solid." However, each of the three classes also includes materials and criteria in addition to those for flammable solids. Therefore, all OSHA flammable solids are NFPA Flammability Hazard 4, 3 or 2 materials, but the converse is not true.

OSHA'S DEFINITION OF "HIGHLY TOXIC"
COMPARED TO DOT, EPA AND NFPA DEFINITIONS

OSHA   DOT

"HIGHLY TOXIC"   =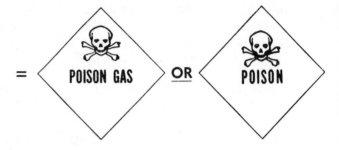

All OSHA highly toxics are DOT Poison A ("Poison Gas" label) or Poison B ("Poison" label) materials, and all DOT Poison A and Poison B materials are OSHA highly toxics.

OSHA   EPA

"HIGHLY TOXIC"   =   "ACUTE HAZARDOUS WASTE"
                    PROBABLE

All OSHA highly toxics are EPA acute hazardous wastes if listed as such by EPA; all EPA acute hazardous wastes are OSHA highly toxics.

OSHA   NFPA

"HIGHLY TOXIC"   =

All OSHA highly toxics are NFPA Health Hazard 4 or Health Hazard 3 materials, but not all NFPA Health Hazard 4 or 3 materials (combustion products) are OSHA highly toxics.

OSHA'S DEFINITION OF "HIGHLY TOXIC"
COMPARED TO DOT, EPA AND NFPA DEFINITIONS

OSHA's definition of "highly toxic" has three components based on tests conducted on laboratory animals. The first is oral toxicity and specifies a median lethal dose (LD 50) of 50 milligrams or less per kilogram of body weight. The second is skin absorption toxicity and specifies an LD 50 of 200 milligrams or less per kilogram of body weight. The third component is inhalation toxicity and specifies a median lethal concentration (LC 50) of 200 parts per million of gas or vapor or 2 milligrams per liter of mist, fume or dust. These three tests are required to determine if a chemical is highly toxic only if adequate data on human toxicity is not available.

DOT's definition of "Poison B" is essentially identical to OSHA's definition of "highly toxic." In addition, DOT has listed certain gases and vapors that are, in very small amounts, dangerous to life and classed them as "Poison A." All OSHA highly toxics are either DOT Poison A or Poison B materials unless the material qualifies for a higher DOT hazard class. All DOT Poison A and Poison B materials are OSHA highly toxics. The DOT label for Poison A material is the "Poison Gas" label and for Poison B material, the "Poison" label.

EPA's definition of "acute hazardous waste" is essentially identical to OSHA's definition of "highly toxic." However, a waste is an acute hazardous waste only if it is listed as such by EPA in its hazardous waste lists. While all OSHA highly toxics meet the criteria for EPA acute hazardous wastes, some may not be EPA acute hazardous wastes because they have not been listed as such by EPA. All EPA acute hazardous wastes are OSHA highly toxics.

NFPA's Health Hazard 4 and Health Hazard 3 cover materials that are included in OSHA's definition of "highly toxic." Health Hazards 4 and 3 also cover materials that give off highly toxic gases and combustion products under fire conditions and Health Hazard 3 includes materials classed as corrosives which are defined separately by OSHA. Therefore, all OSHA highly toxics are NFPA Health Hazard 4 or 3 materials, but not all Health Hazard 4 or 3 materials are OSHA highly toxics.

OSHA'S DEFINITION OF "IRRITANT"
COMPARED TO DOT, EPA AND NFPA DEFINITIONS

OSHA		DOT
"IRRITANT"	=	NO CORRESPONDING DEFINITION

OSHA		EPA
"IRRITANT"	=	NO CORRESPONDING DEFINITION

OSHA		NFPA
"IRRITANT"	=	NO CORRESPONDING DEFINITION

OSHA'S DEFINITION OF "IRRITANT"
COMPARED TO DOT, EPA AND NFPA DEFINITIONS

OSHA's definition of "irritant" focuses on the non-corrosive, reversible, inflammatory effect of a chemical on living tissue. A chemical can be a skin irritant or an eye irritant.

DOT, EPA and **NFPA** do not define hazards that correspond to OSHA's definition of "irritant."

OSHA'S DEFINITION OF "ORGANIC PEROXIDE" COMPARED TO DOT, EPA AND NFPA DEFINITIONS

OSHA   **DOT**

"ORGANIC PEROXIDE"   **=**
PROBABLE

Most OSHA organic peroxides are DOT organic peroxides,
and all DOT organic peroxides are OSHA organic peroxides.

OSHA   **EPA**

"ORGANIC PEROXIDE"   **=**   **"REACTIVE (NORMALLY UNSTABLE)"**
PROBABLE

Most OSHA organic peroxides are EPA reactive (normally unstable) wastes,
but many EPA reactive (normally unstable) wastes are not OSHA organic peroxides.

OSHA   **NFPA**

"ORGANIC PEROXIDE"   **=**   ◇3◇ **OR** ◇2◇ **OR** ◇3/2◇
PROBABLE

Most OSHA organic peroxides are NFPA Flammability Hazard 3 and/or Reactivity Hazard 2 materials,
but many NFPA Flammability Hazard 3 and Reactivity Hazard 2 materials are not OSHA organic proxides.

OSHA'S DEFINITION OF "ORGANIC PEROXIDE"
COMPARED TO DOT, EPA AND NFPA DEFINITIONS

OSHA's definition of "organic peroxide" is a chemistry defintion. Organic peroxides are organic compounds that contain the bivalent -O-O- structure and that are considered to be a structural derivative of hydrogen peroxide where one or both of the hydrogen atoms has been replaced by an organic radical. Organic peroxides are a class of unstable, combustible chemicals.

DOT's definition of "organic peroxide" is essentially identical to OSHA's defintion although it does provide for reclassification of organic peroxides that have another, more predominant hazard. Therefore, most OSHA organic peroxides are DOT organic peroxides and all DOT organic peroxides are OSHA organic peroxides. The DOT label for an organic peroxide is the "Organic Peroxide" label.

EPA does not have a hazardous waste characteristic definition that corresponds closely to OSHA's definition of "organic peroxide." However, since organic peroxides are normally unstable, the EPA reactive (normally unstable) characteristic does correspond to one organic peroxide characteristic. Therefore, many OSHA organic peroxides are EPA reactive (normally unstable) wastes if they become waste. Some of EPA's reactive (normally unstable) wastes are OSHA organic peroxides before they become waste.

NFPA lists "many organic peroxides" as one of the types of materials covered by Flammability Hazard 3. In addition, NFPA's Reactivity Hazard 2 covers materials that are normally unstable. Together, these two hazard classes cover the characteristics of most OSHA organic peroxides although the correspondence is somewhat loose. Most OSHA organic peroxides are NFPA Flammability Hazard 3 and/or NFPA Reactivity Hazard 2 materials. The two NFPA hazard classes include many types of materials other than OSHA organic peroxides.

OSHA'S DEFINITION OF "OXIDIZER"
COMPARED TO DOT, EPA AND NFPA DEFINITIONS

OSHA

DOT

"OXIDIZER"   =

All OSHA oxidizers are DOT oxidizers,
and all DOT oxidizers are OSHA oxidizers.

OSHA

EPA

"OXIDIZER"   =   "OXIDIZER"

All OSHA oxidizers are EPA oxidizers,
and all EPA oxidizers are OSHA oxidizers.

OSHA

NFPA

"OXIDIZER"   =

All OSHA oxidizers are NFPA special hazard "OX" (oxidizer) materials,
and all NFPA special hazard "OX" (oxidizer) materials are OSHA oxidizers.

OSHA'S DEFINITION OF "OXIDIZER"
COMPARED TO DOT, EPA AND NFPA DEFINITIONS

OSHA defines "oxidizer" as a chemical that initiates or promotes combustion in other materials either of itself or through the release of oxygen or other gases.

DOT's definition of "oxidizer" is worded differently than OSHA's definition but the meaning of the two is essentially identical. All OSHA oxidizers are DOT oxidizers unless the material qualifies for a higher DOT hazard class. All DOT oxidizers are OSHA oxidizers. The DOT label for an oxidizer is the "Oxidizer" label.

EPA defines "oxidizer" by reference to DOT's definition of "oxidizer." Therefore, as with DOT, all OSHA oxidizers are EPA oxidizers if they become waste and all EPA oxidizers are OSHA oxidizers before they become waste.

NFPA's Special Hazard "OX" identifies oxidizers. All OSHA oxidizers are NFPA Special Hazard "OX" materials and all NFPA Special Hazard "OX" materials are OSHA oxidizers.

OSHA'S DEFINITION OF "PYROPHORIC"
COMPARED TO DOT, EPA AND NFPA DEFINITIONS

OSHA   **DOT**

"PYROPHORIC"   =
FOR SOLIDS

All OSHA pyrophorics that are solids are DOT flammable solids,
and all DOT flammable solids that are spontaneously combustible are OSHA pyrophorics.

OSHA   **DOT**

"PYROPHORIC"   =
FOR LIQUIDS

All OSHA pyrophorics that are liquids are DOT flammable liquids,
and all DOT flammable liquids that are pyrophorics are OSHA pyrophorics.

Continued . . .

OSHA'S DEFINITION OF "PYROPHORIC"
COMPARED TO DOT, EPA AND NFPA DEFINITIONS

OSHA defines as "pyrophoric" any chemical, solid or liquid, that will spontaneously ignite in air at a temperature of 130°F or below.

DOT divides into two different hazard classes of solid and liquid materials that will spontaneously ignite. DOT's flammable solid hazard class includes spontaneously combustible solid materials. These materials are defined as substances that will spontaneously ignite under conditions normally incident to transportation or upon exposure to air. Since temperatures of up to 130°F can be considered normally incident to transportation, all OSHA pyrophoric solids are DOT flammable solids unless the material also qualifies for a higher DOT hazard class. DOT's flammable liquid hazard class includes pyrophoric liquids which are defined as liquids that will spontaneously ignite in air at or below 130°F. All OSHA pyrophoric liquids are DOT flammable liquids unless the material also qualifies for a higher DOT hazard class. The DOT label for flammable solids is the "Flammable Solid" label and for flammable liquids, the "Flammable Liquid" label.

Continued . . .

OSHA'S DEFINITION OF "PYROPHORIC"
COMPARED TO DOT, EPA AND NFPA DEFINITIONS

OSHA   **EPA**

"PYROPHORIC"   =   **"IGNITABLE SOLID"**
FOR SOLIDS

All OSHA pyrophorics that are solids are EPA ignitable solids,
and all EPA ignitable solids that are spontaneously combustible are OSHA pyrophorics.

OSHA   **EPA**

"PYROPHORIC"   =   **"IGNITABLE LIQUID"**
FOR LIQUIDS

All OSHA pyrophorics that are liquids are EPA ignitable liquids,
and all EPA ignitable liquids that are pyrophoric are OSHA pyrophorics.

Continued . . .

EPA does not have a separate hazard class for pyrophorics. Solids that are spontaneously combustible are included in EPA's definition of "ignitable solid" and liquids that are pyrophoric are included in its definition of "ignitable liquid." Therefore, all OSHA pyrophoric solids are EPA ignitable solids if they become waste and all OSHA pyrophoric liquids are EPA ignitable liquids if they become waste.

Continued . . .

OSHA'S DEFINITION OF "PYROPHORIC"
COMPARED TO DOT, EPA AND NFPA DEFINITIONS

OSHA   NFPA

"PYROPHORIC"   =

All OSHA pyrophorics are NFPA Flammability Hazard 3 materials,
but not all NFPA Flammability Hazard 3 materials are OSHA pyrophorics.

NFPA's Flammability Hazard 3 covers, among other things, all materials which ignite spontaneously when exposed to air. All OSHA pyrophorics are NFPA Flammability Hazard 3 materials, but since Flammability Hazard 3 includes criteria other than spontaneous ignition, not all Flammability Hazard 3 materials are OSHA pyrophorics.

OSHA'S DEFINITION OF "SENSITIZER"
COMPARED TO DOT, EPA AND NFPA DEFINITIONS

OSHA		DOT
"SENSITIZER"	**=**	**NO CORRESPONDING DEFINTION**

OSHA		EPA
"SENSITIZER"	**=**	**NO CORRESPONDING DEFINITION**

OSHA		NFPA
"SENSITIZER"	**=**	**NO CORRESPONDING DEFINITION**

OSHA'S DEFINITION OF "SENSITIZER"
COMPARED TO DOT, EPA AND NFPA DEFINITIONS

OSHA defines "sensitizer" as a chemical that causes an allergic reaction in a substantial proportion of exposed people or animals after repeated exposures.

DOT, EPA and **NFPA** do not define hazard classes that correspond to OSHA's definition of "sensitizer."

OSHA'S DEFINITION OF "TOXIC"
COMPARED TO DOT, EPA AND NFPA DEFINITIONS

OSHA		DOT
"TOXIC"	=	NO CORRESPONDING DEFINITION

OSHA		EPA
"TOXIC"	=	NO CORRESPONDING DEFINITION

OSHA		NFPA
"TOXIC"	=	NO CORRESPONDING DEFINITION

OSHA'S DEFINITION OF "TOXIC"
COMPARED TO DOT, EPA AND NFPA DEFINITIONS

OSHA's definition of "toxic" has three components based on tests conducted on laboratory animals. The first is oral toxicity and specifies a median lethal dose (LD 50) of more than 50 milligrams but not more than 500 milligrams per kilogram of body weight. The second is skin absorption toxicity and specifies an LD 50 of more than 200 milligrams but not more than 1,000 milligrams per kilogram of body weight. The third component is inhalation toxicity and specifies a median lethal concentration (LC 50) of either more than 200 parts per million but not more than 2,000 parts per million of gas or vapor, or more than 2 milligrams per liter but not more than 20 milligrams per liter of mist, fume or dust. These three tests are required to determine if a chemical is toxic only if adequate data on human toxicity is not available.

DOT, **EPA** and **NFPA** do not define hazard classes that correspond to OSHA's definition of "toxic."

OSHA'S DEFINITION OF "UNSTABLE (REACTIVE)" COMPARED TO DOT, EPA AND NFPA DEFINITIONS

OSHA		DOT
"UNSTABLE (REACTIVE)"	**=**	**NO CORRESPONDING DEFINITION**

OSHA		EPA
"UNSTABLE (REACTIVE)"	**=**	**"REACTIVE (NORMALLY UNSTABLE)"**

All OSHA unstable (reactive) chemicals are EPA reactive (normally unstable) wastes, and all EPA reactive (normally unstable) wastes are OSHA unstable (reactive) chemicals.

OSHA		NFPA
"UNSTABLE (REACTIVE)"	**=**	

All OSHA unstable (reactive) chemicals are NFPA Reactivity Hazard 2 materials, but not all NFPA Reactivity Hazard 2 materials (water reactives) are OSHA unstable (reactive) chemicals.

OSHA'S DEFINITION OF "UNSTABLE (REACTIVE)"
COMPARED TO DOT, EPA AND NFPA DEFINITIONS

OSHA defines as "unstable (reactive)" chemicals that will vigorously polymerize, decompose, condense or become self-reactive when subjected to shocks, pressure or temperature. OSHA's definition is a paraphrase of NFPA's definition of "unstable." Chemicals that explode or detonate (cause a sudden, almost instantaneous release of pressure, gas, and heat) when subjected to sudden shock, pressure or high temperature are OSHA explosives.

DOT does not define a hazard class that corresponds to OSHA's definition of "unstable (reactive)."

EPA's "reactive (normally unstable)" characteristic is a paraphrase of NFPA's Reactivity Hazard 2. The term "unstable" in EPA's "reactive (normally unstable)" has the same meaning as in OSHA's definition of "unstable (reactive)" since they are both paraphrases of NFPA Reactivity Hazards. Therefore, all OSHA unstable (reactive) chemicals are EPA reactive (normally unstable) wastes if they become wastes and all EPA reactive (normally unstable) wastes are OSHA unstable (reactive) chemicals before they become wastes.

NFPA's definition of "unstable" is essentially identical to OSHA's definition of "unstable (reactive)" and NFPA's Reactivity Hazard 2 is the hazard class for normally unstable materials. It is also the class for some water reactive materials. Therefore, all OSHA unstable (reactive) chemicals are NFPA Reactivity Hazard 3 materials, but not all materials that have an NFPA Reactivity Hazard 3 are OSHA unstable (reactive) chemicals.

OSHA **DOT**

"WATER REACTIVE"
FOR SOLIDS ONLY =

All OSHA water reactives that are solids are DOT water reactive solids ("Flammable Solid" label plus "Dangerous When Wet" label), and most DOT water reactive solids are OSHA water reactives.

OSHA **EPA**

"WATER REACTIVE" = **"REACTIVE (WATER REACTIVE)"**

All OSHA water reactives are EPA reactive (water reactive) wastes, and all EPA reactive (water reactive) wastes are OSHA water reactives.

OSHA **NFPA**

"WATER REACTIVE" =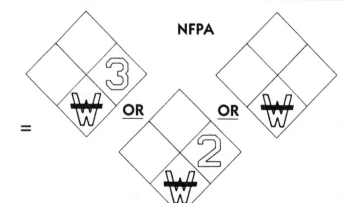

All OSHA water reactives are NFPA Reactivity Hazard 3 or 2 and/or W Special Hazard materials, and most NFPA Reactivity Hazard 3 or 2 and/or W Special Hazard materials are OSHA water reactives.

OSHA'S DEFINITION OF "WATER REACTIVE" COMPARED TO DOT, EPA AND NFPA DEFINITIONS

OSHA defines "water reactive" as a chemical that reacts with water to release a gas that is flammable or a gas that is a health hazard.

DOT's definition of "flammable solid" includes water reactive materials. DOT's definition of "water reactive materials" is confined to solids including sludges and pastes. OSHA's definition of "water reactive" covers water reactive liquids as well as solids. In addition to the generation of flammable or toxic gases, DOT's definition of "water reactive material (solid)" includes materials that become spontaneously flammable when they interact with water. Therefore, most OSHA water reactives that are solids are DOT water reactive materials unless the materials also qualify for a higher DOT hazard class. DOT water reactive materials must carry the "Flammable Solid" label and the "Dangerous When Wet" label. DOT does not have a hazard class for liquid water reactive materials although most of these materials are covered by other hazard classes such as "flammable liquid," "oxidizer" and "corrosive material."

EPA's reactive (water reactive) characteristic for hazardous waste has three components. One component covers wastes that generate toxic gases when mixed with water. The other two components of EPA's definition cover wastes that react violently with water and wastes that form potentially explosive mixtures with water. While the wording of the EPA and the OSHA definitions differ, they both cover the same materials. All OSHA water reactives are EPA reactive (water reactive) wastes if they become wastes and all EPA reactive (water reactive) wastes are OSHA water reactives before they become wastes.

NFPA's Reactivity Hazards 3 and 2 cover materials that react explosively with water and materials that form potentially explosive mixtures with water, respectively. In addition, NFPA's W Special Hazard denotes materials which have an unusual reaction with water. This would cover chemicals that generate health hazard gases. As with EPA reactive (water reactive) wastes, while the wording of the NFPA and the OSHA definitions differ, they both cover the same materials. Therefore, all OSHA water reactives are NFPA Reactivity Hazard 3 or 2 and/or W Special Hazard materials. Since NFPA Reactivity Hazards 3 and 2 also cover hazards other than water reactivity, not all NFPA Reactivity Hazard 3 or 2 materials are OSHA water reactives.

DOT HAZARDOUS MATERIALS

CONTENTS OF CHAPTER

DOT HAZARDOUS MATERIALS

This chapter compares the hazard classification system developed by the Department of Transportation (DOT) for the transportation of hazardous materials (49 CFR Parts 171 through 179) to the systems developed by the Occupational Safety and Health Administration (OSHA) for the Hazard Communication Standard (employee right-to-know regulations), the Environmental Protection Agency (EPA) for the management of hazardous waste and the National Fire Protection Association (NFPA) to identify the fire hazards of materials.

DOT's hazardous materials transportation regulations cover shippers and carriers (transporters) of hazardous materials and transportation by highway, rail, air and water. This chapter focuses on the DOT requirements for highway shipments of hazardous materials although much of the analysis is applicable to the other modes as well. DOT regulates the interstate transportation of hazardous materials by motor vehicle and the inter- and intra-state transportation of hazardous materials that are hazardous wastes regulated by EPA. While DOT does not cover *intra-state* transportation of hazardous materials (except hazardous wastes), many states have adopted regulations that are essentially identical to DOT's. Therefore, the analysis contained in this chapter is also applicable to intra-state transportation of hazardous materials in many states.

The DOT regulations require shippers of hazardous materials (those who offer the material for shipment) to identify which of their materials are hazardous. There are two steps to the identification process. The first is to determine whether or not the material is listed in the DOT's "Hazardous Materials Table" (49 CFR 172.101). For listed materials, the table assigns the proper shipping name, the hazard class and the required label(s). If a material is not listed, the shipper must determine whether it meets the definition of any of 20 DOT hazard classes. The regulations and the "Hazardous Materials Table" specify the proper shipping name and required label(s) for non-listed hazardous materials.

Once a material has been determined to be hazardous and the appropriate hazard class has been identified, the DOT regulations contain requirements for the shipper to prepare a shipping paper for the material, to package the material for shipment, to mark and label the package and to provide placards to the carrier. Shipping papers must include the proper shipping name and hazard class for a material. The regulations specify the labels to be used for particular hazard classes.

Because hazardous material shipping papers and labels identify the hazard class of a material, they are an excellent source of information that can be used to simplify compliance with the OSHA and EPA programs and use of the NFPA labeling system. This chapter is designed to assist the regulated community in using the information on hazardous material shipping papers and/or labels for these purposes.

For most of the hazard classes defined by DOT in the hazardous materials transportation regulations, the chapter identifies the corresponding hazard class defined by OSHA, EPA and NFPA. The chapter also identifies the NFPA fire, reactivity, health and special hazard rating for the class. Three DOT hazard classes, "Radioactive Material," "Etiologic Agent," and "Other Regulated Material (ORM)" are not included in the analysis in this Guide.

Continued...

The hazard class comparisons should only be used for unlisted substances. If a material is listed by DOT in the Hazardous Materials Table (49 CFR 172.101), the proper hazard class and required label(s) are identified in the table. If a waste is listed by EPA in the "F" list (40 CFR 261.31), the "K" list (40 CFR 261.32), the "P" list (40 CFR 261.33) or the "U" list (40 CFR 261.33), it is a hazardous waste with the characteristic and hazardous waste number assigned in the list. If a material is listed in *NFPA 49: Hazardous Chemical Data* or *NFPA 325M: Fire Hazard Properties of Flammable Liquids, Gases, and Volatile Solids,* the fire, reactivity, health and special hazard classes assigned in the lists should be used. OSHA specifies two lists which establish that a chemical is a hazardous chemical, but it still requires the chemical manufacturer to evaluate the chemical to determine its chemical and physical properties and physical and health hazards. This caution about listed materials, therefore, does not apply to OSHA hazardous chemicals.

It is important to note that, of the four hazard classification systems analyzed, only the DOT has a ranking system for determining the appropriate hazard class for a material that meets the criteria of more than one hazard class. This ranking system is used to determine the proper shipping requirements for a material and the proper shipping label(s) for unlisted, multiple hazard class materials. For many multiple hazard class combinations, DOT requires multiple labels. The DOT hazard class ranking system and the regulations covering multiple labeling are presented in the Appendix to this Guide. The comparisons in this chapter can be used to identify materials that are subject to OSHA and EPA regulation and to identify a major hazard represented by the material, but the comparisons do not necessarily identify all of the hazards of a material.

By law and by regulation, all EPA hazardous wastes are DOT hazardous materials. Hazardous materials that meet any of EPA's definitions for characteristic hazardous waste, are hazardous wastes only if they become wastes. For non-wastes, therefore, the analysis in this chapter can be used to determine whether or not the hazardous material is a hazardous waste if it becomes a waste. Wastes that are determined to be hazardous wastes because they have one or more of the hazardous characteristics defined by EPA are assigned hazardous waste numbers based on the characteristic. The hazardous waste number for ignitable wastes (gases, liquids, solids and oxidizers) is D001; for corrosive wastes, D002; and for reactive wastes (cyanide and sulfide bearing wastes, explosives, normally unstable wastes and water reactives) is D003. EPA does not have rules about which characteristic to use if a waste meets the criteria of more than one characteristic. One of the DOT hazard classes not analyzed in this Guide pertains to hazardous wastes. The "ORM-E" hazard class is for hazardous wastes that do not meet the criteria for any other DOT hazard class.

The NFPA diamond is designed to convey information about all of the hazards of a material. The top portion of the diamond is for flammability hazards, the right portion is for reactivity hazards, the left portion is for health hazards and the bottom portion is for special hazards. The degree of hazard is indicated by numbers ranging from 0 to 4, with 4 denoting the highest degree of hazard. The analysis in this chapter identifies the NFPA hazard class and degree(s) that correspond to each DOT hazard class. However, for chemicals that have more than one hazard, the additional hazards must be identified before the proper NFPA label can be completed.

*   *   *

DOT'S DEFINITION OF "BLASTING AGENT"
COMPARED TO OSHA, EPA AND NFPA DEFINITIONS

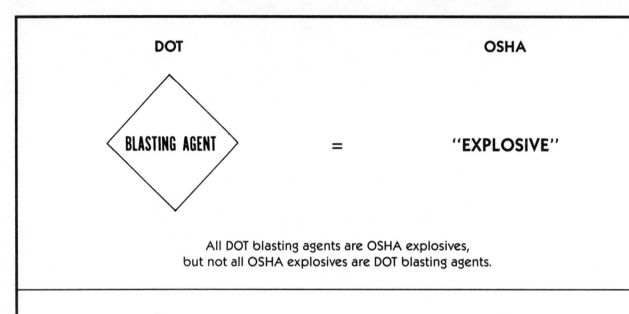

DOT   **OSHA**

BLASTING AGENT   =   "EXPLOSIVE"

All DOT blasting agents are OSHA explosives,
but not all OSHA explosives are DOT blasting agents.

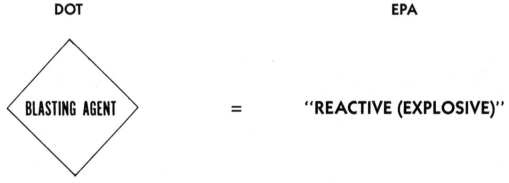

DOT   **EPA**

BLASTING AGENT   =   "REACTIVE (EXPLOSIVE)"

All DOT blasting agents are EPA reactive (explosive) wastes,
but not all EPA reactive (explosive) wastes are DOT blasting agents.

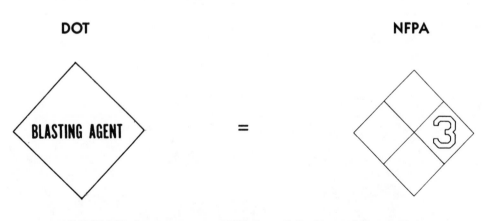

DOT   **NFPA**

BLASTING AGENT   =   3

All DOT blasting agents are NFPA Reactivity Hazard 3 materials,
but not all NFPA Reactivity Hazard 3 materials are DOT blasting agents.

DOT'S DEFINITION OF "BLASTING AGENT"
COMPARED TO OSHA, EPA AND NFPA DEFINITIONS

DOT defines four classes of materials that are commonly referred to as explosives. "Blasting agent" is one of the classes. The other three are Class A, Class B and Class C Explosives. Blasting agents are materials used for blasting and which have a very low potential for ignition during transportation. The DOT label for blasting agents is the "Blasting Agent" label.

OSHA defines "explosive" as a chemical that causes a sudden, almost instantaneous release of pressure, gas and heat when subjected to a sudden shock, pressure or high temperature. Blasting agents, even though they have a very low potential for ignition during transportation, do "explode" with a strong primer. Therefore, all DOT blasting agents are OSHA explosives. OSHA's definition of "explosive" encompasses all explosives, not just DOT blasting agents.

EPA's "reactive (explosive)" characteristic for hazardous waste has three components. Under one of the components, a paraphrase of NFPA's Reactivity 3, a waste is an EPA explosive if it is capable of detonation or explosion when subjected to a strong initiating source or when heated under confinement. This component covers blasting agents. All DOT blasting agents are EPA reactive (explosive) wastes if they become wastes. As with OSHA, EPA's reactive (explosive) characteristic covers all explosives, not just DOT blasting agents.

NFPA's Reactivity Hazard 3 covers materials capable of detonation or explosion when subject to a strong initiating source or when heated under confinement. Reactivity Hazard 3 also includes materials which react explosively with water. All DOT blasting agents are NFPA Reactivity Hazard 3 materials, but because Reactivity Hazard 3 includes water reactives and other explosives, not all Reactivity Hazard 3 materials are DOT blasting agents.

DOT'S DEFINITION OF "COMBUSTIBLE LIQUID"
COMPARED TO OSHA, EPA AND NFPA DEFINITIONS

DOT

OSHA

=

"COMBUSTIBLE LIQUID"

All DOT combustible liquids ("Combustible" placard) are OSHA combustible liquids, and all OSHA combustible liquids are DOT combustible liquids.

DOT

FOR FLASH POINTS BELOW 140°F ONLY

EPA

=

"IGNITABLE LIQUID"

DOT combustible liquids ("Combustible" placard) with a flash point below 140°F are EPA ignitable liquids, and EPA ignitable liquids with a flash point of 100°F or more are DOT combustible liquids.

DOT

NFPA

=

All DOT combustible liquids ("Combustible" placard) are NFPA Flammability Hazard 2 materials, and NFPA Flammability Hazard 2 liquids are DOT combustible liquids.

DOT'S DEFINITION OF "COMBUSTIBLE LIQUID" COMPARED TO OSHA, EPA AND NFPA DEFINITIONS

DOT's definition of "combustible liquid" focuses on the flash point of the liquid. Liquids with a flash point at or above 100 °F and below 200 °F are combustible liquids. DOT does not require a label on packages containing a combustible liquid. However, when combustible liquids are transported in packages with a capacity greater than 110 gallons or in a cargo tank or tank car, a "combustible" placard must be used.

OSHA's definition of "combustible liquid" is essentially identical to DOT's definition. All DOT combustible liquids are OSHA combustible liquids under the Hazard Communication Standard. All OSHA combustible liquids are DOT combustible liquids unless the liquid also qualifies for a higher DOT hazard class.

EPA's definition of "ignitable liquid" also focuses on the flash point of the liquid, but it covers liquids with a flash point less than 140 °F. DOT's definition of "combustible liquid" covers liquids with a flash point between 100 °F and 200 °F. Therefore, only DOT combustible liquids with a flash point less than 140 °F are EPA ignitable liquids if they become waste. Conversely, only EPA ignitable liquids with a flash point between 100 °F and 140 °F are DOT combustible liquids.

NFPA's Flammability Hazard 2 includes liquids with a flash point range that is the same as in DOT's definition of "combustible liquid." Flammability Hazard 2 also includes solids and semi-solids which readily give off flammable vapors. Therefore, all DOT combustible liquids are NFPA Flammability Hazard 2 materials but only liquids with an NFPA Flammability Hazard 2 are DOT combustible liquids.

DOT'S DEFINITION OF "COMPRESSED GAS"
COMPARED TO OSHA, EPA AND NFPA DEFINITIONS

DOT   OSHA

=   "COMPRESSED GAS"

All DOT compressed gases ("Non-Flammable Gas" label) are OSHA compressed gases, and all OSHA compressed gases are DOT compressed gases.

DOT   EPA

=   **NO CORRESPONDING DEFINITION**

DOT   NFPA

=   **NO CORRESPONDING DEFINITION**

DOT'S DEFINITION OF "COMPRESSED GAS"
COMPARED TO OSHA, EPA AND NFPA DEFINITIONS

DOT's definition of "compressed gas" focuses on the pressure of a gas in a container at various temperatures. Compressed gases that are not flammable must be labeled with the DOT "Non-Flammable Gas" label.

OSHA's definition of "compressed gas" is essentially the same as DOT's definition. All DOT compressed gases are OSHA compressed gases and vice versa.

Neither **EPA** nor **NFPA** define hazards that correspond to DOT's definition of "compressed gas."

DOT'S DEFINITION OF "CORROSIVE MATERIAL"
COMPARED TO OSHA, EPA AND NFPA DEFINITIONS

DOT   **OSHA**

=   **"CORROSIVE"**

FOR "SKIN" CORROSIVES ONLY

DOT "skin" corrosive materials are OSHA corrosives and all OSHA corrosives are DOT corrosive materials; but DOT "metal" corrosive materials are not OSHA corrosives.

DOT   **EPA**

=   **"CORROSIVITY"**

FOR "METAL" CORROSIVES ONLY

All DOT "metal" corrosive materials are EPA corrosives, and all EPA "metal" corrosives are DOT corrosive materials.

DOT   **EPA**

=   **"CORROSIVITY"**

PROBABLE

FOR "SKIN" CORROSIVES ONLY

Most DOT "skin" corrosive materials are EPA corrosives, and most EPA "pH" corrosives are DOT corrosive materials.

Continued . . .

DOT'S DEFINITION OF "CORROSIVE MATERIAL" COMPARED TO OSHA, EPA AND NFPA DEFINITIONS

DOT's definition of "corrosive material" has two components: 1) a "skin" component, and 2) a "metal" component. The "skin" component concerns the destructive effect of a substance on human skin. The test specified to determine if a material is "skin" corrosive involves applying the material to the skin of an albino rabbit. The "metal" component concerns the corrosion rate of a substance on metal. The test specified to determine if a material is "metal" corrosive involves applying the material to steel. Containers of materials in the corrosive material hazard class must have the DOT "Corrosive" label.

OSHA's definition of "corrosive" is essentially identical to the "skin" component of DOT's definition of "corrosive material." OSHA does not have a definition in the Hazard Communication Standard that corresponds to the "metal" component of DOT's definition. DOT "skin" corrosive materials are OSHA corrosives but DOT corrosive materials that are "metal" corrosives only are not OSHA corrosives. All OSHA corrosives are DOT corrosive materials.

EPA's definition of "corrosivity" includes a "metal" component and a "pH" component. EPA's "metal" component is identical to DOT's. The "pH" component of EPA's definition specifies that a waste with a pH of 2 or less, or a pH of 12.5 or more is a corrosive hazardous waste. A waste that is corrosive under the "metal" component of EPA's definition is also a DOT corrosive material, regardless of the pH of the waste. Although DOT does not have a definition that corresponds to the "pH" component of EPA's definition, a chemical that has a pH of 2 or less, or a pH of 12.5 or more probably will be classed as corrosive by the "skin" test specified in DOT's definition. However, there may be chemicals with a pH between 2 and 12.5 that are corrosive by the "skin" test, just as there may be chemicals with a pH less than 2 or greater than 12.5 that would not be classed as corrosive by the "skin" test. Therefore, while most DOT "skin" corrosive materials are also EPA "pH" corrosives if they become wastes, there is not necessarily a one-to-one correspondence.

Continued . . .

DOT'S DEFINITION OF "CORROSIVE MATERIAL"
COMPARED TO OSHA, EPA AND NFPA DEFINITIONS

DOT

NFPA

=

FOR "SKIN" CORROSIVES ONLY

All DOT skin corrosive materials are NFPA Health Hazard 3 materials,
but not all NFPA Health Hazard 3 materials are DOT corrosive materials.

DOT'S DEFINITION OF "CORROSIVE MATERIAL" COMPARED TO OSHA, EPA AND NFPA DEFINITIONS

NFPA's Health Hazard 3 includes materials covered by the "skin" component of DOT's definition of "corrosive material." None of NFPA's hazard classes contain a definition that corresponds to the "metal" component of DOT's definition. On the other hand, Health Hazard 3 includes materials that are toxic and those that give off highly toxic combustion products in addition to corrosive materials. Because NFPA does not have a "metal" component in its definition, not all DOT corrosive materials are NFPA Health Hazard 3 materials. Some Health Hazard 3 materials are DOT corrosive materials but not all.

DOT'S DEFINITION OF "EXPLOSIVE, CLASS A" COMPARED TO OSHA, EPA AND NFPA DEFINITIONS

DOT

OSHA

=

"EXPLOSIVE"

All DOT Class A explosives ("Explosive A" label) are OSHA explosives,
but not all OSHA explosives are DOT Class A explosives.

DOT

EPA

=

"REACTIVE (EXPLOSIVE)"

All DOT Class A explosives ("Explosive A" label) are EPA reactive (explosive) wastes,
but not all EPA reactive (explosive) wastes are DOT Class A explosives.

DOT

NFPA

=

All DOT Class A explosives ("Explosive A" label) are NFPA Reactivity Hazard 4 materials,
but not all NFPA Reactivity Hazard 4 materials are DOT Class A explosives.

DOT'S DEFINITION OF "EXPLOSIVE, CLASS A" COMPARED TO OSHA, EPA AND NFPA DEFINITIONS

DOT defines four classes of materials that are commonly referred to as explosives. "Explosive, Class A" is one of the classes. The other three are Class B and Class C Explosives and Blasting Agents. In defining "explosive, Class A," DOT lists nine sequentially numbered types of Class A explosives as well as 13 other types of Class A explosives. Class A explosives are high explosives, detonating explosives and other explosives of maximum hazard. Included in the class are dynamite, lead azide, mercury fulminate, black powder, blasting caps, detonators and detonating primers. The DOT label for Class A explosives is the "Explosive A" label.

OSHA defines "explosive" as a chemical that causes a sudden, almost instantaneous release of pressure, gas and heat when subjected to a sudden shock, pressure or high temperature. All DOT Class A explosives are OSHA explosives. OSHA's definition of "explosive" encompasses all explosives, not just DOT Class A explosives.

EPA's "reactive (explosive)" characteristic for hazardous waste has three components. Under one of the components, DOT Class A and Class B explosives are specifically listed as EPA reactive (explosive) wastes if they become waste.

NFPA's Reactivity Hazard 4 covers materials that are readily capable of detonation or explosion at normal temperatures and pressures. Reactivity Hazard 4 includes materials sensitive to mechanical or localized thermal shock at normal temperature and pressure. The class also includes materials which are readily capable of explosive decomposition. Therefore, all DOT Class A explosives are NFPA Reactivity Hazard 4 materials but not all Reactivity Hazard 4 materials are DOT Class A explosives.

DOT'S DEFINITION OF "EXPLOSIVE, CLASS B" COMPARED TO OSHA, EPA AND NFPA DEFINITIONS

DOT   **OSHA**

   =   **"EXPLOSIVE"**

All DOT Class B explosives ("Explosive B" label) are OSHA explosives,
but not all OSHA explosives are DOT Class B explosives.

DOT   **EPA**

   =   **"REACTIVE (EXPLOSIVE)"**

All DOT Class B explosives ("Explosive B" label) are EPA reactive (explosive) wastes,
but not all EPA reactive (explosive) wastes are DOT Class B explosives.

DOT   **NFPA**

   =   **4**   **OR**   **3**

All DOT Class B explosives ("Explosive B" label) are NFPA Reactivity Hazard 4 or 3 materials,
but not all NFPA Reactivity Hazard 4 or 3 materials are DOT Class B explosives.

DOT'S DEFINITION OF "EXPLOSIVE, CLASS B" COMPARED TO OSHA, EPA AND NFPA DEFINITIONS

DOT defines four classes of materials that are commonly referred to as explosives. "Explosive, Class B" is one of the classes. The other three are Class A and Class C Explosives and Blasting Agents. Class B explosives are materials that function by rapid combustion rather than detonation. The class includes some explosive devices (special fireworks), flash powders, some pyrotechnic signal devices and liquid or solid repellant explosives (some smokeless powders). The DOT label for Class B explosives is the "Explosive B" label.

OSHA defines "explosive" as a chemical that causes a sudden, almost instantaneous release of pressure, gas and heat when subjected to a sudden shock, pressure or high temperature. All DOT Class B explosives are OSHA explosives. OSHA's definition of "explosive" encompasses all explosives, not just DOT Class B explosives.

EPA's "reactive (explosive)" characteristic for hazardous waste has three components. Under one of the components, DOT Class A and Class B explosives are specifically listed as EPA reactive (explosive) wastes if they become waste.

NFPA's Reactivity Hazard 4 covers materials that are readily capable of detonation, explosion or explosive decomposition at normal temperatures and pressures. Reactivity Hazard 4 includes materials sensitive to mechanical or localized thermal shock at normal temperature and pressure. NFPA's Reactivity Hazard 3 covers materials capable of detonation, explosion or explosive decomposition when subject to a strong initiating source or when heated under confinement. Reactivity Hazard 3 also includes materials which react explosively with water. The criteria of Reactivity Hazards 4 and 3 cover DOT Class B explosives. All DOT Class B explosives are NFPA Reactivity Hazard 4 or Reactivity Hazard 3 materials but not all Reactivity Hazard 4 or 3 materials are DOT Class B explosives.

DOT'S DEFINITION OF "EXPLOSIVE, CLASS C"
COMPARED TO OSHA, EPA AND NFPA DEFINITIONS

DOT   **OSHA**

   =   **"EXPLOSIVE"**

All DOT Class C explosives ("Explosive C" label) are OSHA explosives,
but not all OSHA explosives are DOT CLass C explosives.

DOT   **EPA**

   =   **"REACTIVE (EXPLOSIVE)"**

All DOT Class C explosives ("Explosive C" label) are EPA reactive (explosive) wastes,
but not all EPA reactive (explosive) wastes are DOT Class C explosives.

DOT   **NFPA**

   =

All Dot Class C explosives ("Explosive C" label) are NFPA Reactivity Hazard 3 materials,
but not all NFPA Reactivity Hazard 3 materials are DOT Class C explosives.

DOT'S DEFINITION OF "EXPLOSIVE, CLASS C"
COMPARED TO OSHA, EPA AND NFPA DEFINITIONS

DOT defines four classes of materials that are commonly referred to as explosives. "Explosive, Class C" is one of the classes. The other three are Class A and Class B explosives and blasting agents. Class C explosives are manufactured articles which contain restricted amounts of Class A and/or Class B explosives. Class C is the least hazardous of DOT's explosive hazard classes. The DOT label for Class C explosives is the "Explosive C" label.

OSHA defines "explosive" as a chemical that causes a sudden, almost instantaneous release of pressure, gas and heat when subjected to a sudden shock, pressure or high temperature. Class C explosives, even though they are the least hazardous explosives, have the potential to explode or deflagrate when subjected to a sudden shock, pressure or high temperature. Therefore, all DOT Class C explosives are OSHA explosives. OSHA's definition of "explosive" encompasses all explosives, not just DOT Class C explosives.

EPA's "reactive (explosive)" characteristic for hazardous waste has three components. Under one of the components, a paraphrase of NFPA's Reactivity Hazard 3, a waste is an EPA explosive if it is capable of detonation or explosion when subjected to a strong initiating source or when heated under confinement. This component covers Class C explosives. All DOT Class C explosives are EPA reactive (explosive) wastes if they become wastes. As with OSHA, EPA's reactive (explosive) characteristic covers all explosives, not just DOT Class C explosives.

NFPA's Reactivity Hazard 3 covers materials capable of detonation or explosion when subject to a strong initiating source or when heated under confinement. Reactivity Hazard 3 also includes materials which react explosively with water. All DOT Class C explosives are NFPA Reactivity Hazard 3 materials, but because Reactivity Hazard 3 includes water reactives and other explosives, not all Reactivity Hazard 3 materials are DOT Class C explosives.

DOT'S DEFINITION OF "FLAMMABLE COMPRESSED GAS"
COMPARED TO OSHA, EPA AND NFPA DEFINITIONS

DOT

OSHA

= **PARTIAL**

"FLAMMABLE AEROSOL"
OR
"FLAMMABLE GAS"

Most DOT flammable compressed gases ("Flammable Gas" label) are OSHA flammable aerosols or flammable gases; all OSHA flammable aerosols and flammable gases are DOT flammable compressed gases.

DOT

EPA

= **"IGNITABLE COMPRESSED GAS"**

All DOT flammable compressed gases ("Flammable Gas" label) are EPA ignitable compressed gases, and all EPA ignitable compressed gases are DOT flammable compressed gases.

DOT

NFPA

=

All DOT flammable compressed gases ("Flammable Gas" label) are NFPA Flammability Hazard 4 materials, but not all NFPA Flammability Hazard 4 materials are DOT flammable compressed gases.

DOT'S DEFINITION OF "FLAMMABLE COMPRESSED GAS" COMPARED TO OSHA, EPA AND NFPA DEFINITIONS

DOT's definition of "flammable compressed gas" has four components. The first component deals with concentrations at which a gas forms a flammable mixture with air. The "concentration" component has two parts. A gas that satisfies the criterion of either of the parts is a "flammable compressed gas." Under the first part, a gas is a "flammable compressed gas" if it forms a flammable mixture with air when the concentration of the gas is 13 percent or less. Under the second part, a gas is a "flammable compressed gas" if it has a range of flammable concentrations with air that is wider than 12 percent regardless of the lower limit of its range.

The second component to DOT's definition of "flammable compressed gas" is the "flame projection" component. A gas is a "flammable compressed gas" if, using a Bureau of Explosives (BOE) test apparatus, the gas projects a flame more than 18 inches or has a flame that extends back to the valve. The third and fourth components of DOT's definition concern the results of two BOE tests. A gas is a "flammable compressed gas" if, in the BOE open drum test, there is any significant propagation of flame away from the ignition source. A gas is a "flammable compressed gas" if, in the BOE closed drum test, there is any explosion of the vapor-air mixture. The DOT label for flammable compressed gases is the "Flammable Gas" label.

OSHA's definition of "flammable gas" is essentially identical to the "concentration" component of DOT's definition of "flammable compressed gas" and OSHA's definition of "flammable aerosol" is essentially identical to the "flame projection" component of DOT's definition. OSHA does not define hazard criteria that correspond to the other two components of DOT's definition. Therefore, most DOT flammable compressed gases are either OSHA flammable aerosols or OSHA flammable gases. All OSHA flammable aerosols and flammable gases are DOT flammable compressed gases.

EPA's definition of "ignitable compressed gas" simply refers to DOT's definition of "flammable compressed gas." Therefore, the correspondence between DOT's definition and EPA's is one-to-one.

NFPA's Flammability Hazard 4 includes, among other things, flammable gases. All DOT flammable compressed gases are NFPA Flammability Hazard 4 materials. NFPA Flammability Hazard 4, however, also includes other forms of material. Therefore, not all substances with a Flammability Hazard 4 are DOT flammable compressed gases.

DOT'S DEFINITION OF "FLAMMABLE LIQUID"
COMPARED TO OSHA, EPA AND NFPA DEFINITIONS

DOT   OSHA

=   **"FLAMMABLE LIQUID"**

All DOT flammable liquids are OSHA flammable liquids,
and all OSHA flammable liquids are DOT flammable liquids.

DOT   EPA

=   **"IGNITABLE LIQUID"**

All DOT flammable liquids are EPA ignitable liquids,
but not all EPA ignitable liquids (flash points of 100°F or greater) are DOT flammable liquids.

DOT   NFPA

=      **OR**

All DOT flammable liquids are NFPA Flammability Hazard 4 or Flammability Hazard 3 materials,
but not all NFPA Flammability Hazard 4 and 3 materials are DOT flammable liquids.

DOT'S DEFINITION OF "FLAMMABLE LIQUID" COMPARED TO OSHA, EPA AND NFPA DEFINITIONS

DOT's definition of "flammable liquid" focuses on the flash point of a liquid. A liquid with a flash point of less than 100°F is a DOT flammable liquid. DOT flammable liquids must have the DOT "Flammable Liquid" label.

OSHA's definition of "flammable liquid" is essentially identical to DOT's definition. All DOT flammable liquids are OSHA flammable liquids and vice versa.

EPA's definition of "ignitable liquid" includes liquids with flash points of up to 140°F. Therefore, all DOT flammable liquids are EPA ignitable liquids if they become waste, but not all EPA ignitable liquids (those with a flashpoint of 100°F or higher) are DOT flammable liquids.

NFPA's Flammability Hazards 4 and 3 include boiling point criteria in addition to flash point to classify liquids. A liquid with a flash point below 73°F and a boiling point below 100°F is a Flammability Hazard 4. The criteria for Flammability Hazard 3 are:

1) a flash point below 73°F and a boiling point at or above 100°F or

2) a flash point at or above 73°F but below 100°F.

NFPA Flammability Hazards 4 and 3 also cover gases, solids, dusts and mists. Therefore, while all DOT flammable liquids are NFPA Flammability Hazard 4 or Flammability Hazard 3 materials, not all materials classed as NFPA Flammability Hazards 4 or 3 are DOT flammable liquids.

DOT'S DEFINITION OF "FLAMMABLE SOLID"
COMPARED TO OSHA, EPA AND NFPA DEFINITIONS

DOT   OSHA

= "FLAMMABLE SOLID"

All DOT flammable solids ("Flammable Solid" label only) are OSHA flammable solids,
and all OSHA flammable solids are DOT flammable solids.

DOT   EPA

= "IGNITABLE SOLID"

All DOT flammable solids ("Flammable Solid" label only) are EPA ignitable solids,
and all EPA ignitable solids are DOT flammable solids.

DOT   NFPA

All DOT flammable solids ("Flammabile Solid" label only) are NFPA Flammability Hazard 4, 3 or 2 materials,
but not all NFPA Flammability Hazard 4, 3 or 2 materials are DOT flammable solids.

DOT'S DEFINITION OF "FLAMMABLE SOLID" COMPARED TO OSHA, EPA AND NFPA DEFINITIONS

DOT's definition of "flammable solid" has four components. The first is that the solid is likely to cause a fire. This can be the result of friction or retained heat from manufacturing or processing. The second component is that the solid can be ignited readily and when ignited burns very vigorously and persistently. The third component of the DOT definition is that the solid is spontaneously combustible. The final component is that the solid is water reactive. The DOT label for flammable solids is the "Flammable Solid" label. However, for water reactive materials, the "Flammable Solid" label must be accompanied by the "Dangerous When Wet" label. Because DOT water reactives can be distinguished from other flammable solids, they are treated separately in this chapter.

OSHA's definition of "flammable solid" is similar to DOT's without water reactives. Therefore, all DOT flammable solids with only the "Flammable Solid" label are OSHA flammable solids and vice versa.

EPA's definition of "ignitable solid" is essentially identical to OSHA's definition of "flammable solid." All DOT flammable solids with only the "Flammable Solid" label are EPA ignitable solids if they become wastes and all EPA ignitable solids are DOT flammable solids.

NFPA does not have a single hazard class that corresponds closely to DOT's definition of "flammable solid." NFPA Flammability 4 covers materials that will burn readily, including dusts of combustible solids that can form explosive mixtures with air. NFPA Flammability Hazard 3 covers solids that burn very rapidly and those that are spontaneously combustible. NFPA Flammability Hazard 2 covers solids that require moderate heating before they ignite but which readily give off flammable vapors. In combination, the criteria for these three classes cover all of the provisions of DOT's definition of "flammable solid."However, each of the three classes also includes materials and criteria in addition to those for flammable solids. Therefore, all DOT flammable solids with only the "Flammable Solid" label are NFPA Flammability Hazard 4, 3 or 2 materials, but the converse is not true.

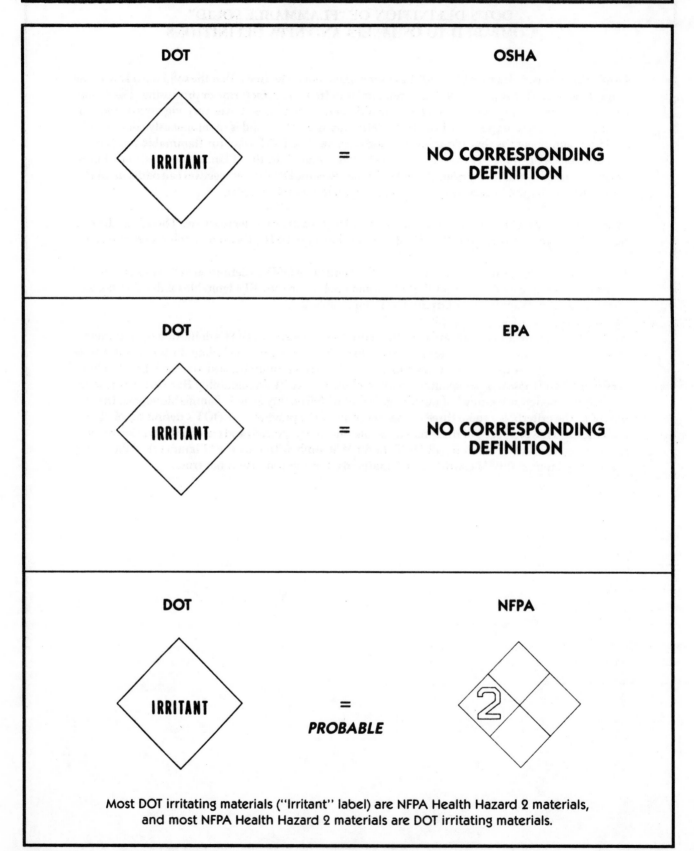

DOT'S DEFINITION OF "IRRITATING MATERIALS"
COMPARED TO OSHA, EPA AND NFPA DEFINITIONS

DOT **OSHA**

IRRITANT = **NO CORRESPONDING DEFINITION**

DOT **EPA**

IRRITANT = **NO CORRESPONDING DEFINITION**

DOT **NFPA**

IRRITANT = *PROBABLE*

Most DOT irritating materials ("Irritant" label) are NFPA Health Hazard 2 materials, and most NFPA Health Hazard 2 materials are DOT irritating materials.

DOT'S DEFINITION OF "IRRITATING MATERIALS" COMPARED TO OSHA, EPA AND NFPA DEFINITIONS

DOT defines "irritating materials" as materials that either under fire conditions or when exposed to air give off dangerous or intensely irritating fumes. The DOT label for irritating materials is the "Irritant" label.

OSHA and **EPA** do not define hazard classes that correspond to DOT's definition of "irritating materials."

NFPA's Health Hazard 2 covers materials that give off toxic or highly irritating combustion products and materials that under fire or normal conditions give off toxic vapors that lack warning properties. Health Hazard 2 covers the portions of DOT's definition on the emission of dangerous or irritating fumes under fire conditions but it does not necessarily cover the emission of such fumes when the material is exposed to air. Therefore, most DOT irritating materials are NFPA Health Hazard 2 materials and most Health Hazard 2 materials are DOT irritating materials, but there is not a one-to-one correspondence.

DOT'S DEFINITION OF "ORGANIC PEROXIDE" COMPARED TO OSHA, EPA AND NFPA DEFINITIONS

DOT **OSHA**

= **"ORGANIC PEROXIDE"**

All DOT organic peroxides are OSHA organic peroxides,
but not all OSHA organic peroxides are DOT organic peroxides.

DOT **EPA**

= **"REACTIVE (NORMALLY UNSTABLE)"**

PROBABLE

Many DOT organic peroxides are EPA reactive (normally unstable) wastes,
but many EPA reactive (normally unstable) wastes are not DOT organic peroxides.

DOT **NFPA**

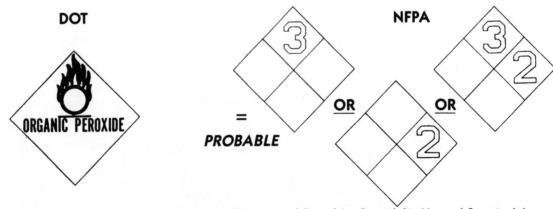

=

PROBABLE

Most DOT organic peroxides are NFPA Flammability Hazard 3 and/or Reactivity Hazard 2 materials,
but many NFPA Flammability Hazard 3 and Reactivity Hazard 2 materials are not DOT organic peroxides.

DOT'S DEFINITION OF "ORGANIC PEROXIDE"
COMPARED TO OSHA, EPA AND NFPA DEFINITIONS

DOT's definition of "organic peroxide" is a chemical definition. Organic peroxides are organic compounds that contain the bivalent -O-O- structure and that are considered to be a structural derivative of hydrogen peroxide where one or both of the hydrogen atoms has been replaced by an organic radical. Organic peroxides are a class of unstable, combustible chemicals. DOT's definition provides for reclassification of organic peroxides that have another hazard that is predominant. The DOT label for an organic peroxide is the "Organic Peroxide" label.

OSHA's definition of "organic peroxide" is essentially identical to DOT's definition except that OSHA's definition does not provide for reclassification of organic peroxides that have another hazard which is predominant. Therefore, all DOT organic peroxides are OSHA organic peroxides but some OSHA organic peroxides may not be DOT organic peroxides.

EPA does not have a hazardous waste characteristic definition that corresponds closely to DOT's definition of "organic peroxide." However, since organic peroxides are normally unstable, the EPA reactive (normally unstable) characteristic does correspond to one organic peroxide characteristic. Therefore, many DOT organic peroxides are EPA reactive (normally unstable) wastes if they become wastes. Some of EPA's reactive (normally unstable) wastes are DOT organic peroxides.

NFPA lists "many organic peroxides" as one of the types of materials covered by Flammability Hazard 3. In addition, NFPA's Reactivity Hazard 2 covers materials that are normally unstable. Together, these two hazard classes cover the characteristics of most DOT organic peroxides although the correspondence is somewhat loose. Most DOT organic peroxides are NFPA Flammability Hazard 3 and/or NFPA Reactivity Hazard 2 materials. The two NFPA hazard classes include many types of materials other than DOT organic peroxides.

DOT'S DEFINITION OF "OXIDIZER"
COMPARED TO OSHA, EPA AND NFPA DEFINITIONS

DOT   **OSHA**

   =   **"OXIDIZER"**

All DOT oxidizers are OSHA oxidizers,
and all OSHA oxidizers are DOT oxidizers.

DOT   **EPA**

   =   **"OXIDIZER"**

All DOT oxidizers are EPA oxidizers,
and all EPA oxidizers are DOT oxidizers.

DOT   **NFPA**

   =

All DOT oxidizers are NFPA special hazard "OX" (oxidizer) materials,
and all NFPA special hazard "OX" (oxidizer) materials are DOT oxidizers.

DOT'S DEFINITION OF "OXIDIZER"
COMPARED TO OSHA, EPA AND NFPA DEFINITIONS

DOT defines "oxidizer" as a material that yields oxygen readily to stimulate combustion of organic matter. The DOT label for an oxidizer is the "Oxidizer" label.

OSHA's definition of "oxidizer" is worded differently than DOT's definition but the meaning of the two is essentially the same. All DOT oxidizers are OSHA oxidizers and vice versa.

EPA defines "oxidizer" by reference to DOT's definition of "oxidizer." Therefore, there is a one-to-one correspondence between DOT's and EPA's definition of oxidizer.

NFPA's Special Hazard "OX" identifies oxidizers. All DOT oxidizers are NFPA Special Hazard "OX" materials and all NFPA Special Hazard "OX" materials are DOT oxidizers.

DOT'S DEFINITION OF "POISON A"
COMPARED TO OSHA, EPA AND NFPA DEFINITIONS

DOT   OSHA

=   "HIGHLY TOXIC"

All DOT Poison A materials ("Poison Gas" label) are OSHA highly toxic materials,
but not all OSHA highly toxics are DOT Poison A materials.

DOT   EPA

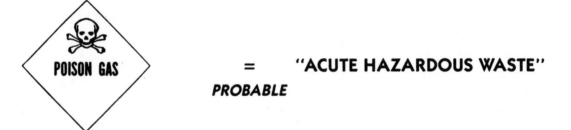

=   "ACUTE HAZARDOUS WASTE"

PROBABLE

DOT Poison A materials ("Poison Gas" label) are EPA acute hazardous wastes if listed as such by EPA,
but not all EPA acute hazardous wastes are DOT Poison A materials.

DOT   NFPA

=

All Dot Poison A materials ("Poison Gas" label) are NFPA Health Hazard 4 materials,
but not all NFPA Health Hazard 4 materials are DOT Poison A materials.

DOT'S DEFINITION OF "POISON A"
COMPARED TO OSHA, EPA AND NFPA DEFINITIONS

DOT defines "Poison A" as gases and vapors that are, in very small amounts, dangerous to life. In addition, DOT lists the materials that are Poison A materials. The DOT label for Poison A material is the "Poison Gas" label.

OSHA's definition of "highly toxic" has three components based on tests conducted on laboratory animals. The first is oral toxicity, which specifies a median lethal dose (LD 50) of 50 milligrams or less per kilogram of body weight. The second is skin absorption toxicity, which specifies an LD 50 of 200 milligrams or less per kilogram of body weight. The third component is inhalation toxicity, which specifies a median lethal concentration (LC 50) of 200 parts per million of gas or vapor or 2 milligrams per liter of mist, fume or dust. These three tests are required to determine if a chemical is highly toxic only if adequate data on human toxicity is not available. All DOT Poison A materials qualify as OSHA highly toxics but the OSHA definition covers materials in addition to DOT Poison A materials.

EPA's definition of "acute hazardous waste" is essentially identical to OSHA's definition of "highly toxic." However, a waste is an acute hazardous waste only if it is listed as such by EPA in its hazardous waste lists. While all DOT Poison A materials meet the criteria for EPA acute hazardous wastes, some may not be EPA acute hazardous wastes because they have not been listed as such by EPA. EPA's definition of "acute hazardous wastes" covers more materials than does DOT's definition of Poison A materials.

NFPA's Health Hazard 4 covers materials that are so toxic that very short exposure could cause death or major residual injury. The class also includes materials that, under fire conditions, give off extremely hazardous fumes. This property of materials is not covered in DOT's definition of Poison A. Therefore, all DOT Poison A materials are NFPA Health Hazard 4 materials but not all Health Hazard 4 materials are DOT Poison A materials.

DOT'S DEFINITION OF "POISON B"
COMPARED TO OSHA, EPA AND NFPA DEFINITIONS

DOT   OSHA

=   "HIGHLY TOXIC"

All DOT Poison B materials ("Poison" label) are OSHA highly toxics,
and all OSHA highly toxics are DOT Poison B materials.

DOT   EPA

=   "ACUTE HAZARDOUS WASTE"

PROBABLE

DOT Poison B materials ("Poison" label) are EPA acute hazardous wastes if listed as such by EPA,
and all EPA acute hazardous wastes are DOT Poison B materials.

DOT   NFPA

=

All DOT Poison B materials ("Poison" label) are NFPA Health Hazard 4 or Health Hazard 3 materials,
but not all NFPA Health Hazard 4 or 3 materials (combustion products) are DOT Poison B materials.

DOT'S DEFINITION OF "POISON B"
COMPARED TO OSHA, EPA AND NFPA DEFINITIONS

DOT's definition of "Poison B" has three components based on tests conducted on laboratory animals. The first is oral toxicity, which specifies a median lethal dose (LD 50) of 50 milligrams or less per kilogram of body weight. The second is skin absorption toxicity, which specifies an LD 50 of 200 milligrams or less per kilogram of body weight. The third component is inhalation toxicity, which specifies a median lethal concentration (LC 50) of 2 milligrams per liter of mist, fume or dust. These three tests are required to determine if a material is a Poison B only if adequate data on human toxicity is not available. DOT Poison A materials meet the criteria of the Poison B definition, but DOT's hierarchical ordering of hazard classes places Poison A materials in a more hazardous class than Poison B materials. The DOT label for Poison B materials is the "Poison" label.

OSHA's definition of "highly toxic" is essentially identical to DOT's definition of "Poison B." All DOT Poison B materials are OSHA highly toxics and all OSHA highly toxics are DOT Poison B materials unless they are Poison A materials.

EPA's definition of "acute hazardous waste" is essentially identical to DOT's definition of "Poison B." However, a waste is an acute hazardous waste only if it is listed as such by EPA in its hazardous waste lists. While all Poison B materials meet the criteria for EPA acute hazardous wastes, some may not be EPA acute hazardous wastes because they have not been listed as such by EPA. All EPA acute hazardous wastes are DOT Poison B materials unless they are Poison A materials.

NFPA's Health Hazard 4 and Health Hazard 3 cover materials that are included in DOT's definition of "Poison B." Health Hazards 4 and 3 also cover materials that give off highly toxic gases and combustion products under fire conditions; Health Hazard 3 also includes materials classed as corrosives which are defined separately by DOT. Therefore, all DOT Poison B materials are NFPA Health Hazard 4 or 3 materials, but not all Health Hazard 4 or 3 materials are DOT Poison B materials.

DOT'S DEFINITION OF "PYROPHORIC LIQUID"
COMPARED TO OSHA, EPA AND NFPA DEFINITIONS

DOT

OSHA

FOR PYROPHORIC LIQUIDS ONLY

=

"PYROPHORIC"

All DOT pyrophoric liquids ("FLammable Liquid" label) are OSHA pyrophorics, and all OSHA pyrophorics that are liquids are DOT pyrophoric liquids.

DOT

EPA

FOR PYROPHORIC LIQUIDS ONLY

=

"IGNITABLE LIQUID"

All DOT pyrophoric liquids ("Flammable Liquid" label) are EPA ignitable liquids, but not all EPA ignitable liquids are DOT pyrophoric liquids.

DOT

NFPA

FOR PYROPHORIC LIQUIDS ONLY

=

All DOT pyrophoric liquids ("Flammable Liquid" label) are NFPA Flammability Hazard 3 materials, but not all NFPA Flammability Hazard 3 materials are DOT pyrophoric liquids.

DOT'S DEFINITION OF "PYROPHORIC LIQUID"
COMPARED TO OSHA, EPA AND NFPA DEFINITIONS

DOT defines as "pyrophoric" any liquid that will spontaneously ignite in dry or moist air at a temperature of 130°F or below. DOT pyrophoric liquids are included in the DOT flammable liquid hazard class. The DOT label for pyrophoric liquids is the "Flammable Liquid" label.

OSHA's definition of "pyrophoric" is essentially identical to DOT's except that OSHA does not restrict its definition to liquids. DOT divides into two different hazard classes spontaneously ignitable solid and liquid materials. Therefore, all DOT pyrophoric liquids are OSHA pyrophorics and all OSHA pyrophorics that are liquids are DOT pyrophoric liquids.

EPA does not have a separate hazard class for pyrophoric liquids. Liquids that are pyrophoric are included in its definition of "ignitable liquid." Therefore, all DOT pyrophoric liquids are EPA ignitable liquids if they become waste. Not all EPA ignitable liquids are DOT pyrophoric liquids.

NFPA's Flammability Hazard 3 covers, among other things, all materials which ignite spontaneously when exposed to air. All DOT pyrophoric liquids are NFPA Flammability Hazard 3 materials, but since Flammability Hazard 3 includes criteria other than spontaneous ignition, not all Flammability Hazard 3 materials are DOT pyrophoric liquids.

DOT'S DEFINITION OF "SPONTANEOUSLY COMBUSTIBLE MATERIAL (SOLID)" COMPARED TO OSHA, EPA AND NFPA DEFINITIONS

DOT

FOR SPONTANEOUSLY COMBUSTIBLES ONLY

OSHA

=

"PYROPHORIC"

All DOT spontaneously combustible materials (solid) ("Flammable Solid" label) are OSHA pyrophorics, and all OSHA pyrophorics that are solids are DOT spontaneously combustible materials (solid).

DOT

FOR SPONTANEOUSLY COMBUSTIBLES ONLY

EPA

=

"IGNITABLE SOLID"

All DOT spontaneously combustible materials (solid) ("Flammable Solid" label) are EPA ignitable solids, but not all EPA ignitable solids are DOT spontaneously combustible materials (solid).

DOT

FOR SPONTANEOUSLY COMBUSTIBLES ONLY

NFPA

=

All DOT spontaneously combustible materials (solid) ("Flammable Solid" label) are NFPA Flammability Hazard 3 materials, but not all Flammability Hazard 3 materials are DOT spontaneously combustible materials (solid).

DOT'S DEFINITION OF "SPONTANEOUSLY COMBUSTIBLE MATERIAL (SOLID)" COMPARED TO OSHA, EPA AND NFPA DEFINITIONS

DOT's flammable solid hazard class includes spontaneously combustible solid materials. These materials are defined as substances that will spontaneously ignite under conditions normally incident to transportation or upon exposure to air. The DOT label for flammable solids is the "Flammable Solid" label.

OSHA defines as "pyrophoric" any chemical, solid or liquid, that will spontaneously ignite in air at a temperature of 130°F or below. Since temperatures of up to 130°F can be considered normally incident to transportation, all DOT spontaneously combustible solid materials are OSHA pyrophorics and all OSHA pyrophorics that are solids are DOT spontaneously combustible solids.

EPA does not have a separate hazard class for spontaneously combustible solid materials. Solids that are spontaneously combustible are included in EPA's definition of "ignitable solid." Therefore, all DOT spontaneously combustible solids are EPA ignitable solids if they become waste.

NFPA's Flammability Hazard 3 covers, among other things, all materials which ignite spontaneously when exposed to air. All DOT spontaneously combustible solid materials are NFPA Flammability Hazard 3 materials, but since Flammability Hazard 3 includes criteria other than spontaneous ignition, not all Flammability Hazard 3 materials are DOT spontaneously combustible solids.

DOT   **OSHA**

PROBABLE

"WATER REACTIVE"

Most DOT water reactive solids ("Flammable Solid" label plus "Dangerous When Wet" label)
are OSHA water reactives;
OSHA water reactives that are solids are DOT water reactive solids.

DOT   **EPA**

PROBABLE

"REACTIVE (WATER REACTIVE)"

Most DOT water reactive solids ("Flammable Solid" label plus "Dangerous When Wet" label),
are EPA reactive (water reactive) wastes,
and all EPA reactive (water reactive) wastes that are solids are DOT water reactive solids.

Continued...

DOT'S DEFINITION OF "WATER REACTIVE MATERIAL (SOLID)" COMPARED TO OSHA, EPA AND NFPA DEFINITIONS

DOT's definition of "flammable solid" includes water reactive materials. DOT's definition of "water reactive materials" is confined to solids, including sludges and pastes. A water reactive solid is a material that reacts with water to give off flammable or toxic gases, or to become spontaneously flammable. DOT water reactive materials must carry the "Flammable Solid" label and the "Dangerous When Wet" label.

OSHA defines "water reactive" as a chemical that reacts with water to release a gas that is flammable or a health hazard. The OSHA hazard class includes water reactive liquids, but does not include chemicals that become spontaneously flammable when they interact with water. Therefore, most DOT water reactive solids are OSHA water reactives and all OSHA water reactives that are solids are DOT water reactive solids.

EPA's reactive (water reactive) characteristic for hazardous waste has three components. One component covers wastes that generate toxic gases when mixed with water. The other two components of EPA's definition cover wastes that react violently with water and wastes that form potentially explosive mixtures with water. The EPA hazard characteristic includes water reactive liquids but does not include chemicals that become spontaneously flammable when they interact with water. Therefore, most DOT water reactive solids are EPA reactive (water reactive) wastes and all EPA reactive (water reactive) wastes that are solids are DOT water reactive solids.

Continued...

DOT'S DEFINITION OF "WATER REACTIVE MATERIAL (SOLID)"
COMPARED TO OSHA, EPA AND NFPA DEFINITIONS

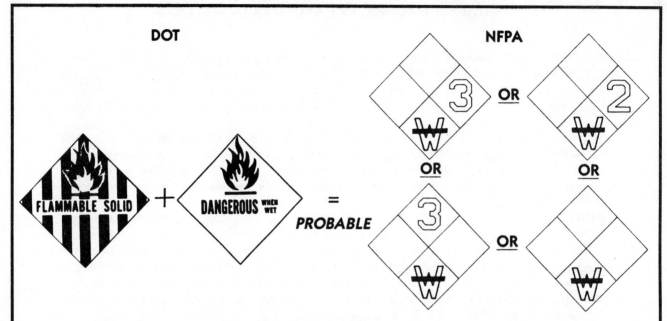

Most DOT water reactive solids ("Flammable Solid" label plus "Dangerous When Wet" label) are NFPA Reactivity Hazard 3 or 2, Flammability Hazard 3 and/or ₩ Special Hazard materials, and most NFPA Reactivity Hazard 3 or 2, Flammability Hazard 3 and/or ₩ Special Hazard materials are DOT water reactive solids.

DOT'S DEFINITION OF "WATER REACTIVE MATERIAL (SOLID)" COMPARED TO OSHA, EPA AND NFPA DEFINITIONS

NFPA's Reactivity Hazards 3 and 2 cover materials that react explosively with water and materials that form potentially explosive mixtures with water, respectively. NFPA Flammability Hazard 3 includes materials that are spontaneously combustible. In addition, NFPA's W Special Hazard denotes materials which have an unusual reaction with water. This would cover chemicals that generate health hazard gases and materials that become spontaneously flammable when they interact with water. Therefore, all DOT water reactive solids are NFPA Reactivity Hazard 3 or 2, Flammability Hazard 3 and/or W Special Hazard materials. Since these NFPA hazard classes also cover hazards other than water reactivity, not all NFPA Reactivity Hazard 3 or 2 materials or Flammability Hazard 3 materials are DOT water reactive solids.

EPA HAZARDOUS WASTES

CONTENTS OF CHAPTER

EPA HAZARDOUS WASTES

This chapter compares the hazard classification system developed by the Environmental Protection Agency for the management of hazardous waste (40 CFR Parts 260 through 271) to the systems developed by the Department of Transportation (DOT) for the transportation of hazardous materials, the Occupational Safety and Health Administration (OSHA) for the Hazard Communication Standard (employee right-to-know regulations) and the National Fire Protection Association (NFPA) to identify the fire hazards of materials.

EPA's hazardous waste regulations cover the management of hazardous waste from "cradle-to-grave," that is, from when the waste is generated, through transportation to treatment, storage and/or disposal of the waste. States are allowed to set up and administer their own hazardous waste management programs as long as their programs are equivalent to or more stringent than EPA's program. Most states have received "authorization" from EPA and do administer their own programs. While some of the states have listed additional hazardous wastes, few have altered EPA's hazardous waste characteristic definitions.

EPA requires waste generators to determine whether or not their wastes are hazardous wastes. There are two steps in the process of identifying hazardous wastes. The first is to determine whether the waste is included in any of EPA's lists of hazardous wastes (the "F" list [40 CFR 261.31], the "K" list [40 CFR 261.32], the "P" list [40 CFR 261.33] or the "U" list [40 CFR 261.33]). The lists assign a hazardous waste number to the waste and indicate the characteristic for which the waste was listed. If a waste is not listed, the generator must determine if it meets any of the hazardous waste characteristics defined by EPA. These characteristics are ignitability, corrosivity, reactivity and EP toxicity. The characteristic of EP toxicity (the "EP" stands for "extraction procedure") is based on the presence and concentration of any of 14 contaminants in a sample of the waste. This characteristic is not analyzed in the Guide.

Each waste characteristic has several components. If a waste meets the criteria of any of the components, it is assigned that characteristic. EPA does not have rules about which characteristic to use if a waste meets the criteria of more than one characteristic. For the analysis in this Guide, the characteristic of ignitability is divided into its four component parts (ignitable gas, ignitable liquid, ignitable solid and oxidizer). The two components of the corrosivity characteristic are analyzed together under the title "corrosive." The eight component parts of the characteristic of reactivity are grouped together into four components that correspond more closely to the hazards defined under the other systems. The four components are "reactive (cyanide and sulfide wastes)," "reactive (explosive)," "reactive (normally unstable)" and "reactive (water reactive)." Each waste characteristic is assigned a hazardous waste number. The number for ignitable wastes (gases, liquids, solids and oxidizers) is D001; for corrosive wastes, D002; and for reactive wastes (cyanide and sulfide bearing wastes, explosives, normally unstable wastes and water reactives) is D003. These numbers must be used by the waste generator in preparing the biennial reports required by EPA.

For each of the component parts of hazard waste characteristics defined by EPA in its hazardous waste management regulations, the chapter identifies the corresponding hazard class defined by DOT, OSHA and NFPA. The chapter also identifies the DOT label appropriate for the hazard class and

Continued...

the NFPA fire, reactivity, health and special hazard rating for the class.

The hazard class comparisons should only be used for unlisted substances. If a material is listed by DOT in the Hazardous Materials Table (49 CFR 172.101), the proper hazard class and required label(s) are identified in the table. If a material is listed in *NFPA 49: Hazardous Chemical Data* or *NFPA 325M: Fire Hazard Properties of Flammable Liquids, Gases, and Volatile Solids,* the fire, reactivity, health and special classes assigned in the lists should be used. OSHA specifies two lists which establish that a chemical is a hazardous chemical, but it still requires the chemical manufacturer to evaluate the chemical to determine its chemical and physical properties and physical and health hazards.

It is important to note that, of the four hazard classification systems analyzed, only the DOT has a ranking system for determining the appropriate hazard class for a material that meets the criteria of more than one hazard class. This ranking system is used to determine the proper shipping requirements for a material and the proper shipping label(s) for unlisted, multiple hazard class materials. For many multiple hazard class combinations, DOT requires multiple labels. The DOT hazard class ranking system and the regulations covering multiple labeling are presented in the Appendix to this Guide. To use the EPA/DOT comparisons in this chapter, identify all of the DOT hazard classes that correspond to the EPA characteristics that have been determined to apply to the waste. If only one DOT hazard class applies, use the hazard class and label indicated in the analysis of the hazard class contained in this chapter. If more than one DOT hazard class applies, consult the Appendix to determine the appropriate hazard class for the material and the label(s) to be used.

Hazardous wastes regulated by EPA are specifically exempted from regulation under the OSHA Hazard Communication Standard. However, the information obtained about a hazardous waste in determining whether it meets any of the waste characteristics can be used to determine whether or not the waste is an OSHA hazardous chemical before it becomes a waste. The analysis in this chapter is designed to assist the regulated community in using information obtained in hazardous waste determinations to identify hazardous chemicals and the characteristic(s) that make them hazardous.

The NFPA diamond is designed to convey information about all of the hazards of a material. The top portion of the diamond is for flammability hazards, the right portion is for reactivity hazards, the left portion is for health hazards and the bottom portion is for special hazards. The degree of hazard is indicated by numbers ranging from 0 to 4, with 4 denoting the highest degree of hazard. The analysis in this chapter identifies the NFPA hazard class and degree(s) that correspond to each component part of the EPA waste characteristics. For wastes that have more than one hazard, the NFPA class and degree should be determined for each hazard. The proper label for the material consists of the appropriate degree of hazard in each portion of the diamond. If more than one degree is identified for a particular NFPA hazard class, use the highest number.

*   *   *

EPA'S DEFINITION OF "ACUTE HAZARDOUS WASTE"
COMPARED TO DOT, OSHA AND NFPA DEFINITIONS

EPA   DOT

"ACUTE HAZARDOUS WASTE"   =   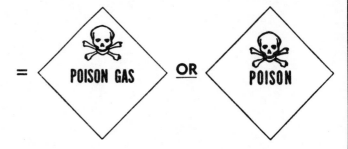   **OR**

All EPA acute hazardous wastes are DOT Poison A ("Poison Gas" label) or Poison B ("Poison" label) materials, but not all DOT Poison A and Poison B materials are EPA acute hazardous wastes.

EPA   OSHA

"ACUTE HAZARDOUS WASTE"   =   "HIGHLY TOXIC"

All EPA acute hazardous wastes are OSHA highly toxic chemicals, but not all OSHA highly toxic chemicals are EPA acute hazardous wastes.

EPA   NFPA

"ACUTE HAZARDOUS WASTE"   =      **OR**

All EPA acute hazardous wastes are NFPA Health Hazard 4 or 3 materials, but not all NFPA Health Hazard 4 and 3 materials are EPA acute hazardous wastes.

EPA'S DEFINITION OF "ACUTE HAZARDOUS WASTE" COMPARED TO DOT, OSHA AND NFPA DEFINITIONS

EPA's definition of "acute hazardous waste" has three components based on tests conducted on laboratory animals. The first is oral toxicity, which specifies a median lethal dose (LD 50) of 50 milligrams or less per kilogram of body weight. The second is skin absorption toxicity, which specifies an LD 50 of 200 milligrams or less per kilogram of body weight. The third component is inhalation toxicity, which specifies a median lethal concentration (LC 50) of 2 milligrams per liter of mist, fume or dust. These three tests are required to determine if a waste is an acute hazardous waste only if adequate data on human toxicity is not available. A waste that meets any of these criteria is an acute hazardous waste if it is listed as such by EPA in its hazardous waste lists.

DOT's definition of "Poison B" is essentially identical to EPA's definition of "acute hazardous waste." In addition, DOT has listed certain gases and vapors that are, in very small amounts, dangerous to life and classed them as "Poison A." All EPA acute hazardous wastes are either DOT Poison A or Poison B materials unless the material qualifies for a higher DOT hazard class. All DOT Poison A and Poison B materials are EPA acute hazardous wastes if they become wastes and if they have been listed as acute hazardous wastes by EPA. The DOT label for Poison A material is the "Poison Gas" label and for Poison B material, the "Poison" label.

OSHA's definition of "highly toxic" is essentially identical to EPA's definition of "acute hazardous waste." All EPA acute hazardous wastes are OSHA highly toxic chemicals before they become wastes. However, OSHA highly toxic chemicals are EPA acute hazardous wastes only if they become wastes and only if they are listed as acute hazardous wastes by EPA.

NFPA's Health Hazard 4 and Health Hazard 3 cover materials that are included in EPA's definition of "acute hazardous waste." Health Hazards 4 and 3 also cover materials that give off highly toxic gases and combustion products under fire conditions and Health Hazard 3 includes materials classed as corrosives which are defined separately by EPA. Therefore, all EPA acute hazardous wastes are NFPA Health Hazard 4 or 3 materials, but not all Health Hazard 4 or 3 materials are EPA acute hazardous wastes.

EPA   **DOT**

"CORROSIVE"   =

FOR "METAL" CORROSIVES ONLY

All EPA "metal" corrosives are DOT corrosive materials,
and all DOT "metal" corrosive materials are EPA corrosives.

EPA   **DOT**

"CORROSIVE"   =

                      PROBABLE

FOR "pH" CORROSIVES ONLY

Most EPA "pH" corrosives are DOT "skin" corrosive materials,
and most DOT "skin" corrosive materials are EPA "pH" corrosives.

Continued . . .

EPA'S DEFINITION OF "CORROSIVE"
COMPARED TO DOT, OSHA AND NFPA DEFINITIONS

EPA's definition of "corrosivity" has two components: (1) a "metal" component, and (2) a "pH" component. The "metal" component concerns the corrosion rate of a substance on metal. The test specified to determine if a waste is corrosive to metal involves applying the waste to steel. The "pH" component specifies that a waste with a pH of 2 or less, or a pH of 12.5 or more is a corrosive hazardous waste.

DOT's definition of "corrosive material" includes a "metal" component and a "skin" component. DOT's "metal" component is identical to EPA's "metal" component. All EPA "metal" corrosives are DOT corrosive materials unless they qualify for a higher DOT hazard class. All DOT "metal" corrosive materials are EPA corrosives if they become wastes. Although EPA does not have a definition that corresponds to the "skin" component of DOT's definition of corrosive material, many materials that are corrosive materials under the DOT's "skin" test have a pH of 2 or less, or a pH of 12.5 or more. However, there are chemicals that are corrosive by the "skin" test that have a pH between 2 and 12.5, just as there may be chemicals that are not corrosive by the "skin" test but that have a pH less than 2 or greater than 12.5. Therefore, most EPA "pH" corrosives are DOT corrosive materials and most DOT "skin" corrosive materials are EPA corrosives.

Continued...

EPA'S DEFINITION OF "CORROSIVE"
COMPARED TO DOT, OSHA AND NFPA DEFINITIONS

EPA   **OSHA**

"CORROSIVE"   =   **"CORROSIVE"**

PROBABLE

FOR "pH" CORROSIVES ONLY

Most EPA "pH" corrosives are OSHA corrosives,
and most OSHA corrosives are EPA corrosives.

EPA   **NFPA**

"CORROSIVE"   =

PROBABLE

FOR "pH" CORROSIVES ONLY

Most EPA "pH" corrosives are NFPA Health Hazard 3 materials;
most NFPA Health Hazard 3 materials that are corrosives are EPA "pH" corrosives.

EPA'S DEFINITION OF "CORROSIVE"
COMPARED TO DOT, OSHA AND NFPA DEFINITIONS

OSHA's definition of "corrosive" is essentially identical to the "skin" component of DOT's definition of "corrosive material." OSHA does not have a definition in the Hazard Communication Standard that corresponds to either the "metal" component or the "pH" component of EPA's definition. Most EPA "pH" corrosives are OSHA corrosives before they become wastes and most OSHA corrosives are EPA corrosives if they become wastes.

NFPA's Health Hazard 3 includes corrosive materials covered by the "skin" component of DOT's definition as well as other materials. None of NFPA's hazard classes contain a definition that corresponds to either the "metal" component or the "pH" component of EPA's definition. Many chemicals that qualify for an NFPA Health Hazard 3 have a pH of 2 or less, or a pH of 12.5 or more. However, there are chemicals with a Health Hazard 3 that have a pH between 2 and 12.5, just as there may be chemicals that do not have a Health Hazard 3 but have a pH less than 2 or greater than 12.5. Therefore, most EPA "pH" corrosives are NFPA Health Hazard 3 materials. Health Hazard 3 also includes materials that are toxic and those that give off highly toxic combustion products as well as corrosive materials. Most Health Hazard 3 materials that are corrosives are EPA corrosives.

EPA'S DEFINITION OF "IGNITABLE COMPRESSED GAS"
COMPARED TO DOT, OSHA AND NFPA DEFINITIONS

EPA   **DOT**

"IGNITABLE COMPRESSED GAS"   =

All EPA ignitable compressed gases are DOT flammable compressed gases ("Flammable Gas" label), and all DOT flammable compressed gases are EPA ignitable compressed gases.

EPA   **OSHA**

"IGNITABLE COMPRESSED GAS"   =   **"FLAMMABLE GAS"**

PROBABLE   **OR**

"FLAMMABLE AEROSOL"

Most EPA ignitable compressed gases are OSHA flammable gases or flammable aerosols; all OSHA flammable gases and flammable aerosols are EPA ignitable compressed gases.

EPA   **NFPA**

"IGNITABLE COMPRESSED GAS"   =

All EPA ignitable compressed gases are NFPA Flammability Hazard 4 materials, but not all NFPA Flammability Hazard 4 materials are EPA ignitable compressed gases.

EPA'S DEFINITION OF "IGNITABLE COMPRESSED GAS" COMPARED TO DOT, OSHA AND NFPA DEFINITIONS

EPA defines "ignitable compressed gas" as a DOT "flammable compressed gas." Therefore, the correspondence between EPA's definition and DOT's is one-to-one.

DOT's definition of "flammable compressed gas" has four components. The first component deals with concentrations at which a gas forms a flammable mixture with air. The "concentration" component has two parts. A gas that satisfies the criterion of either of the parts is a "flammable compressed gas." Under the first part, a gas is a "flammable compressed gas" if it forms a flammable mixture with air when the concentration of the gas is 13 percent or less. Under the second part, a gas is a "flammable compressed gas" if it has a range of flammable concentrations with air that is wider than 12 percent regardless of the lower limit of its range.

The second component to DOT's definition of "flammable compressed gas" is the "flame projection" component. A gas is a "flammable compressed gas" if, using a Bureau of Explosives (BOE) test apparatus, the gas projects a flame more than 18 inches or has a flame that extends back to the valve. The third and fourth components of DOT's definition concern the results of two BOE tests. A gas is a "flammable compressed gas" if, in the BOE open drum test, there is any significant propagation of flame away from the ignition source. A gas is a "flammable compressed gas" if, in the BOE closed drum test, there is any explosion of the vapor-air mixture. The DOT label for flammable compressed gases is the "Flammable Gas" label.

OSHA's definition of "flammable gas" is essentially identical to the "concentration" component of DOT's definition of "flammable compressed gas" and OSHA's definition of "flammable aerosol" is essentially identical to the "flame projection" component of DOT's definition. OSHA does not define hazard criteria that correspond to the other two components of DOT's definition. Therefore, most EPA ignitable compressed gases are either OSHA flammable aerosols or OSHA flammable gases before they become wastes. All OSHA flammable aerosols and flammable gases are EPA ignitable compressed gases if they become wastes.

NFPA's Flammability Hazard 4 includes, among other things, flammable gases. All EPA ignitable compressed gases are NFPA Flammability Hazard 4 materials. NFPA Flammability Hazard 4, however, also includes other forms of material. Therefore, not all substances with a Flammability Hazard 4 are EPA ignitable compressed gases.

EPA   **DOT**

"IGNITABLE LIQUID" =

*FOR FLASH POINTS
BELOW 100ºF ONLY*

EPA ignitable liquids with flash points below 100°F are DOT flammable liquids;
all DOT flammable liquids are EPA ignitable liquids.

EPA   **DOT**

"IGNITABLE LIQUID" =

*FOR FLASH POINTS
100ºF OR HIGHER ONLY*

EPA ignitable liquids with flash points of 100°F or higher
are DOT combustible liquids ("Combustible" placard);
DOT combustible liquids with flash points below 140°F are EPA ignitable liquids.

Continued . . .

EPA'S DEFINITION OF "IGNITABLE LIQUID" COMPARED TO DOT, OSHA AND NFPA DEFINITIONS

EPA's definition of "ignitable liquid" focuses on the flash point of a liquid. A liquid with a flash point of up to 140°F is an "ignitable liquid."

DOT's definition of "flammable liquid" includes liquids with flash points of less than 100°F and its definition of "combustible liquid" includes liquids with flash points at or above 100°F but below 200°F. Unless a liquid also qualifies for a higher DOT hazard class, EPA ignitable liquids with flash points below 100°F are DOT flammable liquids, and EPA ignitable liquids with flash points at or above 100°F are DOT combustible liquids. DOT flammable liquids and DOT combustible liquids with flash points lower than 140°F are EPA ignitable liquids if they become wastes. DOT flammable liquids must have the DOT "Flammable Liquid" label. DOT does not require a label on packages containing a combustible liquid. However, when combustible liquids are transported in packages with a capacity of greater than 110 gallons or in a cargo tank or tank car, a "Combustible" placard must be used.

While DOT's definitions for "flammable liquid" and "combustible liquid" do not have a flash point of 140°F as a criterion, the hazard classification system developed by the International Maritime Organization (IMO) does use a flash point of 140°F as the upper boundary for Class 3.3 Flammable Liquids. The class includes liquids with flash points of between 75°F and 140°F. The upper flash point limit for IMO Class 3.3 Flammable Liquids corresponds to the upper flash point limit for EPA ignitable liquids. The IMO Class for a variety of materials is indicated in the Optional Hazardous Materials Table (49 CFR 172.102).

Continued . . .

EPA'S DEFINITION OF "IGNITABLE LIQUID"
COMPARED TO DOT, OSHA AND NFPA DEFINITIONS

EPA   OSHA

"IGNITABLE LIQUID"   =   "FLAMMABLE LIQUID"

FOR FLASH POINTS
BELOW 100°F ONLY

EPA ignitable liquids with flash points below 100°F are OSHA flammable liquids;
all OSHA flammable liquids are EPA ignitable liquids.

EPA   OSHA

"IGNITABLE LIQUID"   =   "COMBUSTIBLE LIQUID"

FOR FLASH POINTS
100°F OR HIGHER ONLY

EPA ignitable liquids with flash points of 100°F or higher are OSHA combustible liquids;
OSHA combustible liquids with flash points below 140°F are EPA ignitable liquids.

Continued...

OSHA's definitions of "flammable liquid" and "combustible liquid" are essentially the same as DOT's definitions. EPA's "ignitable liquid" has the same correspondence to OSHA's "flammable liquid" and "combustible liquid" as it does to DOT's two definitions.

Continued...

EPA'S DEFINITION OF "IGNITABLE LIQUID"
COMPARED TO DOT, OSHA AND NFPA DEFINITIONS

EPA   **NFPA**

"IGNITABLE LIQUID"   =   [diamond with 4]   **OR**   [diamond with 3]

FOR FLASH POINTS
BELOW 100ºF ONLY

EPA ignitable liquids with flash points below 100°F are NFPA Flammability Hazard 4 or 3 materials;
NFPA Flammability Hazard 4 or 3 materials that are liquids are EPA ignitable liquids.

EPA   **NFPA**

"IGNITABLE LIQUID"   =   [diamond with 2]

FOR FLASH POINTS
100ºF OR HIGHER ONLY

EPA ignitable liquids with flash points of 100°F or higher are NFPA Flammability Hazard 2 materials;
NFPA Flammability Hazard 2 materials that are liquids with flash points below 140°F
are EPA ignitable liquids.

NFPA's Flammability Hazards 4 and 3 include boiling point criteria in addition to flash point to classify liquids. A liquid with a flash point below 73 °F and a boiling point below 100 °F is a Flammability Hazard 4. The criteria for Flammability Hazard 3 are:

1) a flash point below 73 °F and a boiling point at or above 100 °F, or

2) a flash point at or above 73 °F but below 100 °F.

NFPA Flammability Hazards 4 and 3 also cover gases, solids, dusts and mists. NFPA's Flammability Hazard 2 includes liquids with a flash point between 100 °F and 200 °F, the same range as DOT's "combustible liquid." Flammability Hazard 2 also includes solids and semi-solids which readily give off flammable vapors. EPA ignitable liquids with flash points below 100 °F are NFPA Flammability Hazard 4 or 3 materials and EPA ignitable liquids with flash points of 100 °F or higher are NFPA Flammability Hazard 2 materials. NFPA Flammability Hazard 4 and 3 materials that are liquids and NFPA Flammability Hazard 2 materials that are liquids with flash points lower than 140 °F are EPA ignitable liquids.

EPA'S DEFINITION OF "IGNITABLE SOLID"
COMPARED TO DOT, OSHA AND NFPA DEFINITIONS

EPA   DOT

"IGNITABLE SOLID"   =

All EPA ignitable solids are DOT flammable solids,
but not all DOT flammable solids (water reactives) are EPA ignitable solids.

EPA   OSHA

"IGNITABLE SOLID"   =   "FLAMMABLE SOLID"

All EPA ignitable solids are OSHA flammable solids,
and all OSHA flammable solids are EPA ignitable solids.

EPA   NFPA

"IGNITABLE SOLID"   =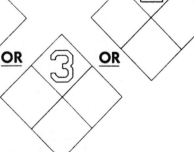

All EPA ignitable solids are NFPA Flammability Hazard 4, 3 or 2 materials,
but not all NFPA Flammability Hazard 4, 3 or 2 materials are EPA ignitable solids.

EPA'S DEFINITION OF "IGNITABLE SOLID" COMPARED TO DOT, OSHA AND NFPA DEFINITIONS

EPA's definition of "ignitable solid" has two components. The first is that the solid is likely to cause a fire. This can be the result of friction, absorption of moisture or spontaneous chemical change. The second component is that when the solid is ignited it burns very vigorously and persistently.

DOT's definition of "flammable solid" is similar to EPA's definition of "ignitable solid," but DOT includes water reactive material in its definition while EPA does not. Therefore, all EPA ignitable solids are DOT flammable solids unless the material qualifies for a higher DOT hazard class. Not all DOT flammable solids (water reactives) are EPA ignitable solids. The DOT label for a flammable solid is the "Flammable Solid" label.

OSHA's definition of "flammable solid" is essentially identical to EPA's definition of "ignitable solid." All EPA ignitable solids are OSHA flammable solids before they become wastes and all OSHA flammable solids are EPA ignitable solids if they become wastes.

NFPA does not have a single hazard class that corresponds closely to EPA's definition of "ignitable solid." NFPA Flammability 4 covers materials that will burn readily, including dusts of combustible solids that can form explosive mixtures with air. NFPA Flammability Hazard 3 covers solids that burn very rapidly and those that are spontaneously combustible. NFPA Flammability Hazard 2 covers solids that require moderate heating before they ignite but which give off flammable vapors. In combination, the criteria for these three classes cover all of the provisions of EPA's definition of "ignitable solid." However, each of the three classes also includes materials and criteria in addition to those for ignitable solids. Therefore, all EPA ignitable solids are NFPA Flammability Hazard 4, 3 or 2 materials, but the converse is not true.

EPA'S DEFINITION OF "OXIDIZER"
COMPARED TO DOT, OSHA AND NFPA DEFINITIONS

EPA   DOT

"OXIDIZER"   =

All EPA oxidizers are DOT oxidizers,
and all DOT oxidizers are EPA oxidizers.

EPA   OSHA

"OXIDIZER"   =   "OXIDIZER"

All EPA oxidizers are OSHA oxidizers,
and all OSHA oxidizers are EPA oxidizers.

EPA   NFPA

"OXIDIZER"   =

All EPA oxidizers are NFPA Special Hazard "OX" (oxidizer) materials,
and all NFPA Special Hazard "OX" (oxidizer) materials are EPA oxidizers.

EPA'S DEFINITION OF "OXIDIZER"
COMPARED TO DOT, OSHA AND NFPA DEFINITIONS

EPA defines "oxidizer" as a DOT "oxidizer." Therefore, there is a one-to-one correspondence between EPA's and DOT's definition of oxidizer.

DOT defines "oxidizer" as a material that yields oxygen readily to stimulate combustion of organic matter. The DOT label for an oxidizer is the "Oxidizer" label.

OSHA's definition of "oxidizer" is worded differently than DOT's definition but the meaning of the two is essentially the same. All EPA oxidizers are OSHA oxidizers before they become wastes and all OSHA oxidizers are EPA oxidizers if they become wastes.

NFPA's Special Hazard "OX" identifies oxidizers. All EPA oxidizers are NFPA Special Hazard "OX" materials and all NFPA Special Hazard "OX" materials are EPA oxidizers.

EPA'S DEFINITION OF "REACTIVE (CYANIDE AND SULFIDE WASTES)" COMPARED TO DOT, OSHA AND NFPA DEFINITIONS

EPA		DOT
"REACTIVE (CYANIDE AND SULFIDE WASTES)"	=	NO CORRESPONDING DEFINITION

EPA		OSHA
"REACTIVE (CYANIDE AND SULFIDE WASTES)"	=	NO CORRESPONDING DEFINITION

EPA		NFPA
"REACTIVE (CYANIDE AND SULFIDE WASTES)"	=	NO CORRESPONDING DEFINITION

EPA'S DEFINITION OF "REACTIVE (CYANIDE AND SULFIDE WASTES)" COMPARED TO DOT, OSHA AND NFPA DEFINITIONS

EPA's hazardous waste characteristic of "reactive (cyanide and sulfide wastes)" covers cyanide or sulfide bearing wastes which can generate dangerous quantities of toxic gases when exposed to pH conditions between 2 and 12.5.

DOT, OSHA and **NFPA** do not have hazard classes that correspond to EPA's "reactive (cyanide and sulfide wastes)" hazardous waste characteristic.

EPA'S DEFINITION OF "REACTIVE (EXPLOSIVE)"
COMPARED TO DOT, OSHA AND NFPA DEFINITIONS

EPA

DOT

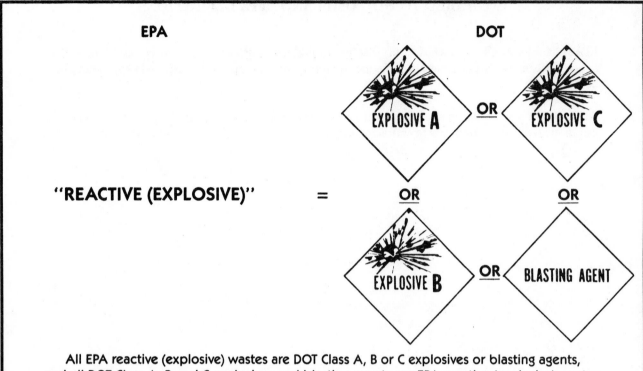

"REACTIVE (EXPLOSIVE)" = EXPLOSIVE **A** **OR** EXPLOSIVE **C**

OR **OR**

EXPLOSIVE **B** **OR** BLASTING AGENT

All EPA reactive (explosive) wastes are DOT Class A, B or C explosives or blasting agents, and all DOT Class A, B and C explosives and blasting agents are EPA reactive (explosive) wastes.

Continued . . .

EPA'S DEFINITION OF "REACTIVE (EXPLOSIVE)" COMPARED TO DOT, OSHA AND NFPA DEFINITIONS

EPA's reactive (explosive) characteristic for hazardous waste has three components. First, DOT Class A and Class B explosives are EPA explosives. The second component of EPA's reactive (explosive) characteristic is a paraphrase of NFPA's Reactivity 4. A waste is an EPA explosive if it is readily capable of detonation or explosion at standard temperature and pressure. Under the third component, a paraphrase of NFPA's Reactivity 3, a waste is an EPA explosive if it is capable of detonation or explosion when subjected to a strong initiating source or when heated under confinement.

DOT defines three classes of explosives (Classes A, B and C) and a class of blasting agents. Class A explosives are high explosives, detonating explosives and other explosives of maximum hazard. Class B explosives are deflagrating explosives or explosives that react by rapid burning rather than detonating. Class C explosives are manufactured articles that contain restricted quantities of Class A or Class B explosives. Blasting agents are materials used for blasting and have a very low potential for ignition during transportation. DOT Class A and Class B explosives are covered by definition in EPA's "reactive (explosive)." Class C explosives and blasting agents are capable of detonation, explosive reaction or explosive decomposition if subjected to a strong initiating source or if heated. Therefore, all EPA reactive (explosive) wastes are DOT Class A, Class B or Class C explosives or blasting agents. Conversely, all DOT explosives (Class A, B and C explosives and blasting agents) are EPA reactive (explosive) wastes. The DOT labels required for the four classes are "Explosive A," "Explosive B," "Explosive C" and "Blasting Agent" respectively.

EPA'S DEFINITION OF "REACTIVE (EXPLOSIVE)"
COMPARED TO DOT, OSHA AND NFPA DEFINITIONS

EPA		OSHA
"REACTIVE (EXPLOSIVE)"	**=**	**"EXPLOSIVE"**

All EPA reactive (explosive) wastes are OSHA explosives,
and all OSHA explosives are EPA reactive (explosive) wastes.

EPA		NFPA
"REACTIVE (EXPLOSIVE)"	**=**	 **OR** 

All EPA reactive (explosive) wastes are NFPA Reactivity Hazard 4 or 3 materials,
and all NFPA Reactivity Hazard 4 materials are EPA reactive (explosive) wastes,
but not all NFPA Reactivity Hazard 3 materials (water reactives) are EPA reactive (explosive) wastes.

OSHA's definition of "explosive" is relatively simple. A chemical that causes a sudden, almost instantaneous release of pressure, gas and heat when subjected to a sudden shock, pressure or high temperature is an explosive. The criteria of this definition are covered by EPA's definition of "reactive (explosive)." All EPA reactive (explosive) wastes are OSHA explosives before they become wastes and all OSHA explosives are EPA reactive (explosive) wastes if they become wastes.

NFPA's Reactivity Hazard 4 covers materials that are readily capable of detonation or explosion at normal temperatures and pressures. Reactivity Hazard 4 includes materials sensitive to mechanical or localized thermal shock at normal temperature and pressure. NFPA's Reactivity Hazard 3 covers materials capable of detonation or explosion when subject to a strong initiating source or when heated under confinement. Reactivity Hazard 3 also includes materials which react explosively with water. Two of the components of EPA's definition of "reactive (explosive)" are paraphrases of the explosive portion of NFPA's Reactivity Hazard 4 and 3. All EPA reactive (explosive) wastes are NFPA Reactivity Hazard 4 or 3 materials and all NFPA Reactivity Hazard 4 materials are EPA reactive (explosive) wastes. Because NFPA Reactivity Hazard 3 also includes water reactives, not all materials in the class are EPA reactive (explosive) wastes.

EPA'S DEFINITION OF "REACTIVE (NORMALLY UNSTABLE)" COMPARED TO DOT, OSHA AND NFPA DEFINITIONS

EPA

DOT

"REACTIVE (NORMALLY UNSTABLE)" = **NO CORRESPONDING DEFINITION**

EPA

OSHA

"REACTIVE (NORMALLY UNSTABLE)" = **"UNSTABLE (REACTIVE)"**

All EPA reactive (normally unstable) wastes are OSHA unstable (reactive) chemicals, and all OSHA unstable (reactive) chemicals are EPA reactive (normally unstable) wastes.

EPA

NFPA

"REACTIVE (NORMALLY UNSTABLE)" =

All EPA reactive (normally unstable) wastes are NFPA Reactivity Hazard 2 materials, but not all NFPA Reactivity Hazard 2 materials (water reactives) are EPA reactive (normally unstable) wastes.

EPA'S DEFINITION OF "REACTIVE (NORMALLY UNSTABLE)" COMPARED TO DOT, OSHA AND NFPA DEFINITIONS

EPA's "reactive (normally unstable)" characteristic is a paraphrase of NFPA's Reactivity Hazard 2. Reactive (normally unstable) waste is waste that is normally unstable and readily undergoes violent change without detonating. NFPA defines "unstable" materials as those which vigorously polymerize, decompose or condense or become self-reactive and undergo other violent chemical changes.

DOT does not define a hazard class that corresponds to OSHA's definition of "unstable (reactive)."

OSHA defines as "unstable (reactive)" chemicals that will vigorously polymerize, decompose, condense or become self-reactive when subjected to shocks, pressure or temperature. OSHA's definition is a paraphrase of NFPA's definition of "unstable." The term "unstable" in OSHA's definition of "unstable (reactive)" has the same meaning as in EPA's "reactive (normally unstable)." Therefore, all EPA reactive (normally unstable) wastes are OSHA unstable (reactive) chemicals before they become wastes and all OSHA unstable (reactive) chemicals are EPA reactive (normally unstable) wastes if they become wastes.

NFPA's Reactivity Hazard 2 is paraphrased by EPA for its definition of "reactive (normally unstable)." Reactivity Hazard 2 is also the class for some water reactive materials. Therefore, all EPA reactive (normally unstable) wastes are NFPA Reactivity Hazard 2 materials, but not all NFPA Reactivity Hazard 2 materials (water reactives) are EPA reactive (normally unstable) wastes.

EPA'S DEFINITION OF "REACTIVE (WATER REACTIVE)" COMPARED TO DOT, OSHA AND NFPA DEFINITIONS

EPA   DOT

"REACTIVE (WATER REACTIVE)" =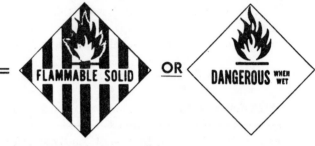
FOR SOLIDS ONLY

All EPA reactive (water reactive) wastes that are solids
are DOT water reactives ("Flammable Solid" label plus "Dangerous When Wet" label),
but not all DOT water reactives are EPA reactive (water reactive) wastes.

EPA   OSHA

"REACTIVE (WATER REACTIVE)"   =   **"WATER REACTIVE"**

All EPA reactive (water reactive) wastes are OSHA water reactives,
and all OSHA water reactives are EPA reactive (water reactive) wastes.

EPA   NFPA

"REACTIVE (WATER REACTIVE)"   =

All EPA reactive (water reactive) wastes
are NFPA Reactivity Hazard 3 or 2 and/or W Special Hazard materials, and most NFPA
Reactivity Hazard 3 or 2 and/or W Special Hazard materials are EPA reactive (water reactive) wastes.

EPA'S DEFINITION OF "REACTIVE (WATER REACTIVE)" COMPARED TO DOT, OSHA AND NFPA DEFINITIONS

EPA's reactive (water reactive) characteristic for hazardous waste has three components. One component covers wastes that generate toxic gases when mixed with water. The other two components of EPA's definition cover wastes that react violently with water and wastes that form potentially explosive mixtures with water. These latter two components are paraphrases of the water reactive portion of NFPA Reactivity Hazard 2.

DOT's definition of "flammable solid" includes water reactive materials. DOT's definition of "water reactive materials" is confined to solids, including sludges and pastes. EPA's definition of "reactive (water reactive)" covers water reactive liquids as well as solids. In addition to the generation of flammable or toxic gases, DOT's definition of "water reactive material (solid)" includes materials that become spontaneously flammable when they interact with water. Therefore, most EPA reactive (water reactive) wastes that are solids are DOT water reactive materials unless the materials also qualify for a higher DOT hazard class. Except for materials that become spontaneously flammable when they interact with water, DOT water reactive materials are EPA reactive (water reactive) wastes if they become wastes. DOT water reactive materials must carry the "Flammable Solid" label and the "Dangerous When Wet" label. DOT does not have a hazard class for liquid water reactive materials although most of these materials are covered by other hazard classes such as "flammable liquid," "oxidizer" and "corrosive material."

OSHA defines "water reactive" as a chemical that reacts with water to release a gas that is flammable or a health hazard. While the wording of the EPA and the OSHA definitions differ, they both cover the same materials. All EPA reactive (water reactive) wastes are OSHA water reactives before they become wastes and all OSHA water reactives are EPA reactive (water reactive) wastes if they become wastes.

NFPA's Reactivity Hazards 3 and 2 cover materials that react explosively with water and materials that form potentially explosive mixtures with water, respectively. EPA's definition of "reactive (water reactive)" is a paraphrase of the water reactive portion of NFPA's Reactivity Hazard 2. In addition to these two hazard classes, NFPA's W Special Hazard denotes materials which have an unusual reaction with water. This would cover chemicals that generate health hazard gases. Therefore, all EPA reactive (water reactive) wastes are NFPA Reactivity Hazard 3 or 2 and/or W Special Hazard materials. Since these NFPA hazard classes also cover hazards other than water reactivity, not all NFPA Reactivity Hazard 3 or 2 materials are EPA reactive (water reactive) wastes.

EPA'S DEFINITION OF "TOXIC WASTE"
COMPARED TO DOT, OSHA AND NFPA DEFINITIONS

EPA		DOT
"TOXIC WASTE"	=	NO CORRESPONDING DEFINITION

EPA		OSHA
"TOXIC WASTE"	=	NO CORRESPONDING DEFINITION

EPA		NFPA
"TOXIC WASTE"	=	NO CORRESPONDING DEFINITION

EPA'S DEFINITION OF "TOXIC WASTE"
COMPARED TO DOT, OSHA AND NFPA DEFINITIONS

EPA defines "toxic waste" as a waste that contains any of the toxic substances listed in "Appendix VIII - Hazardous Constituents" to "Part 261 - Identification and Listing of Hazardous Waste" of its regulations. Substances are to be listed in Appendix VIII if scientific studies have shown them to be toxic, carcinogens, mutagens or teratogens. If a substance listed in Appendix VIII is present in a waste, the waste must be listed by EPA as a hazardous waste unless EPA determines that the waste does not pose a hazard to human health or the environment if improperly managed.

DOT, OSHA and **NFPA** do not have hazard classes that correspond to EPA's definition of "toxic waste."

CHAPTER 4

TEXTS OF HAZARD DEFINITIONS

CONTENTS OF CHAPTER

ENVIRONMENTAL PROTECTION AGENCY

40 CFR 261.11   Criteria for listing hazardous waste.

(a) The Administrator shall list a solid waste as a hazardous waste only upon determining that the solid waste meets one of the following criteria:

(2) It has been found to be fatal to humans in low doses or, in the absence of data on human toxicity, it has been shown in studies to have an oral LD 50 toxicity (rat) of less than 50 milligrams per kilogram, an inhalation LC 50 toxicity (rat) of less than 2 milligrams per liter, or a dermal LD 50 toxicity (rabbit) of less than 200 milligrams per kilogram or is otherwise capable of causing or significantly contributing to an increase in serious irreversible, or incapacitating reversible, illness. (Waste listed in accordance with these criteria will be designated **Acute Hazardous Waste.**)

OCCUPATIONAL SAFETY AND HEALTH ADMINISTRATION

Appendix A to 29 CFR 1910.1200   Health Hazard Definitions (Mandatory)

3. **Highly Toxic:**   A chemical falling within any of the following categories:

(a) A chemical that has a median lethal dose (LD_{50}) of 50 milligrams or less per kilogram of body weight when administered orally to albino rates weighing between 200 and 300 grams each.

(b) A chemical that has a median lethal dose (LD_{50}) of 200 milligrams or less per kilogram of body weight when administered by continuous contact for 24 hours (or less if death occurs within 24 hours) with the bare skin of albino rabbits weighing between two and three kilograms each.

(c) A chemical that has a median lethal concentration (LC_{50}) in air of 200 parts per million by volume or less of gas or vapor, or 2 milligrams per liter or less of mist, fume, or dust, when administered by continuous inhalation for one hour (or less if death occurs within one hour) to albino rats weighing between 200 and 300 grams each.

DEPARTMENT OF TRANSPORTATION

49 CFR 173.326   Poison A.

(a) For the purpose of Parts 170-189 of this subchapter, extremely dangerous poisons, Class A are poisonous gases or liquids of such nature that a very small amount of the gas, or vapor of the liquid, mixed with air is dangerous to life. This class includes the following:

    (1) Bromacetone.
    (2) Cyanogen.
    (3) Cyanogen chloride containing less than 0.9 percent water.
    (4) Diphosgene.
    (5) Ethyldichlorarsine.
    (6) Hydrocyanic acid.
    (7) [Reserved]
    (8) Methyldichlorarsine.
    (9) [Reserved]
    (10) Nitrogen peroxide (tetroxide).
    (11) [Reserved]
    (12) Phosgene (disphosgene).
    (13) Nitrogen tetroxide-nitric oxide mixtures containing up to 33.2 percent weight nitric oxide.

49 CFR 173.343   Poison B.

(a) For the purposes of Parts 170-189 of this subchapter and except as otherwise provided in this part, Class B poisons are those substances, liquid or solid (including pastes and semisolids), other than Class A poisons or Irritating materials, which are known to be so toxic to man as to afford a hazard to health during transportation; or which, in the absence of adequate data on human toxicity, are presumed to be toxic to man because they fall within any one of the following categories when tested on laboratory animals:

(1) Oral toxicity.   Those which produce death within 48 hours in half or more than half of a group of 10 or more white laboratory rats weighing 200 to 300 grams at a single dose of 50 milligrams or less per kilogram of body weight, when administered orally.

(2) Toxicity on inhalation.   Those which produce death within 48 hours in half or more than half of a group of 10 or more white laboratory rats weighing 200 to 300 grams, when inhaled continuously for a period of one hour or less at a concentration of 2 milligrams or less per liter of vapor, mist, or dust, provided such concentration is likely to be encountered by man when the chemical product is used in any reasonable forseeable manner.

(3) Toxicity by skin absorption.   Those which produce death within 48 hours in half or more than half of a group of 10 or more rabbits tested at a dosage of 200 milligrams or less per kilogram body weight, when administered by continuous contact with the bare skin for 24 hours or less.

NATIONAL FIRE PROTECTION ASSOCIATION

NFPA 704   Identification of the Fire Hazards of Materials

Chapter 2 Health Hazards

4   Materials which on very short exposure could cause death or major residual injury even though prompt medical treatment were given, including those which are too dangerous to be approached without specialized protective equipment. This degree should include:

Materials which can penetrate ordinary rubber protective clothing;

Materials which under normal conditions or under fire conditions give off gases which are extremely hazardous (i.e., toxic or corrosive) through inhalation or through contact with or absorption through the skin.

3   Materials which on short exposure could cause serious temporary or residual injury even though prompt medical treatment were given, including those requiring protection from all bodily contact. This degree should include:

Materials giving off highly toxic combustion products;

Materials corrosive to living tissue or toxic by skin absorption.

DEPARTMENT OF TRANSPORTATION

40 CFR 173.114a   **Blasting agents.**

(a) Definition of a blasting agent.   A blasting agent is a material designed for blasting which has been tested in accordance with paragraph (b) of this section and found to be so insensitive that there is very little probability of accidental initiation to explosion or of transition from deflagration to detonation.

OCCUPATIONAL SAFETY AND HEALTH ADMINISTRATION

29 CFR 1910.1200   Hazard Communication.

(c) Definitions.

"Explosive" means a chemical that causes a sudden, almost instantaneous release of pressure, gas, and heat when subjected to sudden shock, pressure, or high temperature.

ENVIRONMENTAL PROTECTION AGENCY

40 CFR 261.23   Characteristic of reactivity. [*Reactive (Explosive)*]

(a) A solid waste exhibits the characteristic of reactivity if a representative sample of the waste has *any* of the following properties:

(6) It is capable of detonation or explosive reaction if it is subjected to a strong initiating source or if heated under confinement.

(7) It is readily capable of detonation or explosive decomposition or reaction at standard temperature and pressure.

(8) It is a forbidden exlosive as defined in 49 CFR 173.51, or a Class A Explosive as defined in 49 CFR 173.53 or a Class B Explosive as defined in 49 CFR 173.88.

NATIONAL FIRE PROTECTION ASSOCIATION

NFPA 704   Identification of the Fire Hazards of Materials

Chapter 4 Reactivity (Instability) Hazards

3   Materials which in themselves are capable of detonation or explosive decomposition or explosive reaction but which require a strong initiating source or which must be heated under confinement before initiation. This degree should include materials which are sensitive to thermal or mechanical shock at elevated temperatures and pressures or which react explosively with water without requiring heat or confinement.

DEFINITIONS OF "CARCINOGEN" AND CORRESPONDING TERMS

OCCUPATIONAL SAFETY AND HEALTH ADMINISTRATION

Appendix A to 29 CFR 1910.1200   Health Hazard Definitions (Mandatory)

1. **Carcinogen:**   A chemical is considered to be a carcinogen if:

   (a) It has been evaluated by the International Agency for Research on Cancer (IARC), and found to be a carcinogen or potential carcinogen; or

   (b) It is listed as a carcinogen or potential carcinogen in the *Annual Report on Carcinogens* published by the National Toxicology Program (NTP) (latest edition); or,

   (c) It is regulated by OSHA as a carcinogen.

ENVIRONMENTAL PROTECTION AGENCY

40 CFR 261.11   Criteria for listing hazardous waste.

   (a) The Administrator shall list a solid waste as a hazardous waste only upon determining that the solid waste meets one of the following criteria:

   (3) It contains any of the toxic constituents listed in Appendix VIII...

   Substances will be listed on Appendix VIII only if they have been shown in scientific studies to have toxic, carcinogenic, mutagenic or teratogenic effects on humans or other life forms. (Wastes listed in accordance with these criteria will be designated **Toxic Wastes.**)

DEPARTMENT OF TRANSPORTATION

49 CFR 173.115   Flammable, combustible, and pyrophoric liquids; definitions.

(b) **Combustible liquid.**

(1) For the purposes of this subchapter, a combustible liquid is defined as any liquid that does not meet the definition of any other classification specified in this subchapter and has a flash point at or above 100 °F (37.8 °C) and below 200 °F (93.3 °C) except any mixture having one component or more with a flash point at 200 °F (93.3 °C) or higher, that makes up at least 99 percent of the total volume of the mixture.

(2) For the purposes of this subchapter:

(i) An aqueous solution containing 24 percent or less alcohol by volume is considered to have a flash point of no less than 100 °F (37.8 °C) if the remainder of the solution is not subject to this subchapter, and

(ii) An aqueous solution containing 24 percent or less alcohol by volume is not subject to the requirements of this subchapter if it contains no less than 50 percent water and no material (other than the alcohol) which is subject to this subchapter.

OCCUPATIONAL SAFETY AND HEALTH ADMINISTRATION

29 CFR 1910.1200   Hazard Communication.

(c) Definitions.

"**Combustible liquid**" means any liquid having a flashpoint at or above 100 °F (37.8 °C) but below 200 °F (93.3 °C), except any mixture having components with flashpoints of 200 °F (93.3 °C), or higher, the total volume of which make up 99 percent or more of the total volume of the mixture.

NATIONAL FIRE PROTECTION ASSOCIATION

NFPA 704   Identification of the Fire Hazards of Materials

Chapter 3 Flammability Hazards

2   Materials that must be moderately heated or exposed to relatively high ambient temperatures before ignition can occur. Materials in this degree would not under normal conditions form hazardous atmospheres with air, but under high ambient temperatures or under moderate heating may release vapor in sufficient quantities to produce hazardous atmospheres with air. This degree should include:

Liquids having a flash point above 100 °F (37.8 °C), but not exceeding 200 °F (93.4 °C);

Solids and semisolids which readily give off flammable vapors.

ENVIRONMENTAL PROTECTION AGENCY

40 CFR 261.21   Characteristic of **ignitability.**

(a) A solid waste exhibits the characteristic of ignitability if a representative sample of the waste has any of the following properties:

(1) It is a **liquid,** other than an aqueous solution containing less than 24 percent alcohol by volume and has flash point less than 60 °C (140 °F), as determined by a Pensky-Martens Closed Cup Tester, using the test method specified in ASTM Standard D-93-79 or D-93-80 (incorporated by reference, see Section 260.11), or a Setaflash Closed Cup Tester, using the test method specified in ASTM Standard D-3278-78 (incorporated by reference, see Section 260.11), or as determined by an equivalent test method approved by the Administrator under procedures set forth in Sections 260.20 and 260.21.

DEFINITIONS OF "COMPRESSED GAS" AND CORRESPONDING TERMS

OCCUPATIONAL SAFETY AND HEALTH ADMINISTRATION

29 CFR 1910.1200   Hazard Communication.

(c) Definitions.

"**Compressed gas**" means:

(i) A gas or mixture of gases having, in a container, an absolute pressure exceeding 40 psi at 70°F (21.1°C); or

(ii) A gas or mixture of gases having, in a container, an absolute pressure exceeding 104 psi at 130°F (54.4°C) regardless of the pressure at 70°F (21.1°C); or

(iii) A liquid having a vapor pressure exceeding 40 psi at 100°F (37.8°C) as determined by ASTM D-323-72.

DEPARTMENT OF TRANSPORTATION

49 CFR 173.300   Definitions.

(a) **Compressed gas.**   The term "compressed gas" shall designate any material or mixture having in the container an absolute pressure exceeding 40 p.s.i. at 70°F or, regardless of the pressure at 70°F, having an absolute pressure exceeding 104 p.s.i. at 130°F; or any liquid flammable material having a vapor pressure exceeding 40 p.s.i. absolute at 100°F as determined by ASTM Test D-323.

DEPARTMENT OF TRANSPORTATION

49 CFR 173.240   **Corrosive material;** definition.

(a) For the purpose of this subchapter, a corrosive material is a liquid or solid that causes visible destruction or irreversible alterations in human skin tissue at the site of contact, or in the case of leakage from its packaging, a liquid that has a severe corrosion rate on steel.

 (1) A material is considered to be destructive or to cause irreversible alteration in human skin tissue if when tested on the intact skin of the albino rabbit by the technique described in Appendix A to this part, the structure of the tissue at the site of contact is destroyed or changed irreversibly after an exposure period of 4 hours or less.

 (2) A liquid is considered to have a severe corrosion rate if its corrosion rate exceeds 0.250 inch per year (IPY) on steel (SAE 1020) at a test temperature of 130°F. An acceptable test is described in NACE Standard TM-01-69.

ENVIRONMENTAL PROTECTION AGENCY

40 CFR 261.22   Characteristic of **corrosivity.**

(a) A solid waste exhibits the characteristic of corrosivity if a representative sample of the waste has either of the following properties:

 (1) It is aqueous and has a pH less than or equal to 2 or greater than or equal to 12.5, as determined by a pH meter using either an EPA test method or an equivalent test method approved by the Administrator under the procedures set forth in Sections 260.20 and 260.21. The EPA test method for pH is specified as Method 5.2 in "Test Methods for the Evaluation of Solid Waste, Physical/Chemical Methods" (incorporated by reference, see Section 260.11).

 (2) It is a liquid and corrodes steel (SAE 1020) at a rate greater than 6.35 mm (0.250 inch) per year at a test temperature of 55°C (130°F) as determined by the test method specified in NACE (National Association of Corrosion Engineers) Standard TM-01-69 as standardized in "Test Methods for the Evaluation of Solid Waste, Physical/Chemical Methods" (incorporated by reference, see Section 260.11) or an equivalent test method approved by the Administrator under the procedures set forth in Sections 260.20 and 260.21.

OCCUPATIONAL SAFETY AND HEALTH ADMINISTRATION

Appendix A to 29 CFR 1910.1200   Health Hazard Definitions (Mandatory)

2. **Corrosive:**   A chemical that causes visible destruction of, or irreversible alterations in, living tissue by chemical action at the site of contact. For example, a chemical is considered to be corrosive if, when tested on the intact skin of albino rabbits by the method described by the U.S. Department of Transportation in Appendix A to 49 CFR Part 173, it destroys or changes irreversibly the structure of the tissue at the site of contact following an exposure period of four hours. This term shall not refer to action on inanimate surfaces.

NATIONAL FIRE PROTECTION ASSOCIATION

NFPA 704   Identification of the Fire Hazards of Materials

Chapter 2 Health Hazards

3   Materials which on short exposure could cause serious temporary or residual injury even though prompt medical treatment were given, including those requiring protection from all bodily contact. This degree should include:

Materials giving off highly toxic combustion products;

Materials corrosive to living tissue or toxic by skin absorption.

DEFINITIONS OF "EXPLOSIVE" AND CORRESPONDING TERMS

OCCUPATIONAL SAFETY AND HEALTH ADMINISTRATION

29 CFR 1910.1200   Hazard Communication

(c) Definitions.

"**Explosive**" means a chemical that causes a sudden, almost instantaneous release of pressure, gas, and heat when subjected to sudden shock, pressure, or high temperature.

DEPARTMENT OF TRANSPORTATION

49 CFR 173.53   Definition of **Class A explosives.**

[*This section defines nine numbered types of Class A explosives as well as thirteen other types of explosives. All of the types are high explosives, detonating explosives or other explosives of maximum hazard and include dynamite, lead azide, mercury fulminate, black powder, blasting caps, detonators and detonating primers.*]

49 CFR 173.88   Definition of **Class B explosives.**

(a) Explosives, Class B, are defined as those explosives which in general function by rapid combustion rather than detonation and include some explosive devices such as special fireworks, flash powders, some pyrotechnic signal devices and liquid or solid propellant explosives which include some smokeless powders. These explosives are further specifically described in paragraphs (b) to (g) of this section.

49 CFR 173.100   Definition of **Class C explosives.**

(a) Explosives, Class C, are defined as certain types of manufactured articles which contain Class A, or Class B explosives, or both, as components in restricted quantities, and certain types of fireworks. These explosives are further specifically described in this section.

49 CFR 173.114a   **Blasting agents.**

(a) Definition of a blasting agent.   A blasting agent is a material designed for blasting which has been tested in accordance with paragraph (b) of this section and found to be so insensitive that there is very little probability of accidental initiation to explosion or of transition from deflagration to detonation.

ENVIRONMENTAL PROTECTION AGENCY

40 CFR 261.23   Characteristic of reactivity. [*Reactive (Explosive)*]

(a) A solid waste exhibits the characteristic of reactivity if a representative sample of the waste has *any* of the following properties:

    (6) It is capable of detonation or explosive reaction if it is subjected to a strong initiating source or if heated under confinement.

    (7) It is readily capable of detonation or explosive decomposition or reaction at standard temperature and pressure.

    (8) It is a forbidden explosive as defined in 49 CFR 173.51, or a Class A Explosive as defined in 49 CFR 173.53 or a Class B Explosive as defined in 49 CFR 173.88.

NATIONAL FIRE PROTECTION ASSOCIATION

NFPA 704   Identification of the Fire Hazards of Materials

Chapter 4 Reactivity (Instability) Hazards

4 Materials which in themselves are readily capable of detonation or of explosive decomposition or explosive reaction at normal temperatures and pressures. This degree should include materials which are sensitive to mechanical or localized thermal shock at normal temperatures and pressures.

3 Materials which in themselves are capable of detonation or explosive decomposition or explosive reaction but which require a strong initiating source or which must be heated under confinement before initiation. This degree should include materials which are sensitive to thermal or mechanical shock at elevated temperatures and pressures or which react explosively with water without requiring heat or confinement.

DEPARTMENT OF TRANSPORTATION

49 CFR 173.53   Definition of **Class A explosives.**

[*This section defines nine numbered types of Class A explosives as well as thirteen other types of explosives. All of the types are high explosives, detonating explosives or other explosives of maximum hazard and include dynamite, lead azide, mercury fulminate, black powder, blasting caps, detonators and detonating primers.*]

OCCUPATIONAL SAFETY AND HEALTH ADMINISTRATION

29 CFR 1910.1200   Hazard Communication.

(c) Definitions.

"**Explosive**" means a chemical that causes a sudden, almost instantaneous release of pressure, gas, and heat when subjected to sudden shock, pressure, or high temperature.

ENVIRONMENTAL PROTECTION AGENCY

40 CFR 261.23   Characteristic of reactivity. [*Reactive (Explosive)*]

(a) A solid waste exhibits the characteristic of reactivity if a representative sample of the waste has *any* of the following properties:

(6) It is capable of detonation or explosive reaction if it is subjected to a strong initiating source or if heated under confinement.

(7) It is readily capable of detonation or explosive decomposition or reaction at standard temperature and pressure.

(8) It is a forbidden explosive as defined in 49 CFR 173.51, or a Class A Explosive as defined in 49 CFR 173.53 or a Class B Explosive as defined in 49 CFR 173.88.

NATIONAL FIRE PROTECTION ASSOCIATION

NFPA 704   Identification of the Fire Hazards of Materials

   Chapter 4   Reactivity (Instability) Hazards

     4   Materials which in themselves are readily capable of detonation or of explosive decomposition or explosive reaction at normal temperatures and pressures. This degree should include materials which are sensitive to mechanical or localized thermal shock at normal temperatures and pressures.

DEPARTMENT OF TRANSPORTATION

49 CFR 173.88   Definition of **Class B explosives.**

(a) Explosives, Class B, are defined as those explosives which in general function by rapid combustion rather than detonation and include some explosive devices such as special fireworks, flash powders, some pyrotechnic signal devices and liquid or solid propellant explosives which include some smokeless powders. These explosives are further specifically described in paragraphs (b) to (g) of this section.

OCCUPATIONAL SAFETY AND HEALTH ADMINISTRATION

29 CFR 1910.1200   Hazard Communication.

(c) Definitions.

"Explosive" means a chemical that causes a sudden, almost instantaneous release of pressure, gas, and heat when subjected to sudden shock, pressure, or high temperature.

ENVIRONMENTAL PROTECTION AGENCY

40 CFR 261.23   Characteristic of reactivity. [*Reactive (Explosive)*]

(a) A solid waste exhibits the characteristic of reactivity if a representative sample of the waste has *any* of the following properties:

(6) It is capable of detonation or explosive reaction if it is subjected to a strong initiating source or if heated under confinement.

(7) It is readily capable of detonation or explosive decomposition or reaction at standard temperature and pressure.

(8) It is a forbidden explosive as defined in 49 CFR 173.51, or a Class A Explosive as defined in 49 CFR 173.53 or a Class B Explosive as defined in 49 CFR 173.88.

NATIONAL FIRE PROTECTION ASSOCIATION

NFPA 704   Identification of the Fire Hazards of Materials

Chapter 4   Reactivity (Instability) Hazards

4   Materials which in themselves are readily capable of detonation or of explosive decomposition or explosive reaction at normal temperatures and pressures. This degree should include materials which are sensitive to mechanical or localized thermal shock at normal temperatures and pressures.

3   Materials which in themselves are capable of detonation or explosive decomposition or explosive reaction but which require a strong initiating source or which must be heated under confinement before initiation. This degree should include materials which are sensitive to thermal or mechanical shock at elevated temperatures and pressures or which react explosively with water without requiring heat or confinement.

DEFINITIONS OF "EXPLOSIVE, CLASS C" AND CORRESPONDING TERMS

DEPARTMENT OF TRANSPORTATION

49 CFR 173.100   Definition of **Class C explosives.**

(a) Explosives, Class C, are defined as certain types of manufactured articles which contain Class A, or Class B explosives, or both, as components but in restricted quantities, and certain types of fireworks. These explosives are further specifically described in this section.

OCCUPATIONAL SAFETY AND HEALTH ADMINISTRATION

29 CFR 1910.1200   Hazard Communication.

(c) Definitions.

"Explosive" means a chemical that causes a sudden, almost instantaneous release of pressure, gas, and heat when subjected to sudden shock, pressure, or high temperature.

ENVIRONMENTAL PROTECTION AGENCY

40 CFR 261.23   Characteristic of reactivity. [*Reactive (Explosive)*]

(a) A solid waste exhibits the characteristic of reactivity if a representative sample of the waste has *any* of the following properties:

(6) It is capable of detonation or explosive reaction if it is subjected to a strong initiating source or if heated under confinement.

(7) It is readily capable of detonation or explosive decomposition or reaction at standard temperature and pressure.

(8) It is a forbidden explosive as defined in 49 CFR 173.51, or a Class A Explosive as defined in 49 CFR 173.53 or a Class B Explosive as defined in 49 CFR 173.88.

NATIONAL FIRE PROTECTION ASSOCIATION

NFPA 704   Identification of the Fire Hazards of Materials

Chapter 4   Reactivity (Instability) Hazards

3   Materials which in themselves are readily capable of detonation or explosive decomposition or explosive reaction but which require a strong initiating source or which must be heated under confinement before initiation. This degree should include materials which are sensitive to thermal or mechanical shock at elevated temperatures and pressures or which react explosively with water without requiring heat or confinement.

DEFINITIONS OF "FLAMMABLE AEROSOL"
AND CORRESPONDING TERMS

OCCUPATIONAL SAFETY AND HEALTH ADMINISTRATION

29 CFR 1910.1200   Hazard Communication.

(c) Definitions.

"Flammable" means a chemical that falls into one of the following categories:

(i) "Aerosol, flammable" means an aerosol that, when tested by the method described in 16 CFR 1500.45 [*Method for determining extremely flammable and flammable contents of self-pressurized containers*], yields a flame projection exceeding 18 inches at full valve opening, or a flashback (a flame extending back to the valve) at any degree of valve opening;

DEPARTMENT OF TRANPSPORTATION

49 CFR 173.300   Definitions.

(b) **Flammable compressed gas.** Any compressed gas as defined in paragraph (a) of this section shall be classed as "flammable gas" if any one of the following occurs:

(1) Either a mixture of 13 percent or less (by volume) with air forms a flammable mixture or the flammable range with air is wider than 12 percent regardless of the lower limit. These limits shall be determined at atmospheric temperature and pressure. The method of sampling and test procedure shall be acceptable to the Bureau of Explosives and approved by the Director, OHMT.

(2) Using the Bureau of Explosives' Flame Projection Apparatus, the flame projects more than 18 inches beyond the ignition source with valve opened fully, or, the flame flashes back and burns at the valve with any degree of valve opening.

(3) Using the Bureau of Explosives' Open Drum Apparatus, there is any significant propagation of flame away from the ignition source.

(4) Using the Bureau of Explosives' Closed Drum Apparatus, there is any explosion of the vapor-air mixture in the drum.

ENVIRONMENTAL PROTECTION AGENCY

40 CFR 261.21   Characteristic of ignitability.

(a) A solid waste exhibits the characteristic of ignitability if a representative sample of the waste has any of the following properties:

(3) It is an **ignitable compressed gas** as defined in 49 CFR 173.300 and as determined by the test methods described in that regulation or equivalent test methods approved by the Administrator under Sections 260.20 and 260.21.

NATIONAL FIRE PROTECTION ASSOCIATION

NFPA 704   Identification of the Fire Hazards of Materials

Chapter 3 Flammability Hazards

4   Materials which will rapidly or completely vaporize at atmospheric pressure and normal ambient temperature or which are readily dispersed in air, and which will burn readily. This degree should include:

Gases;

Cryogenic materials;

Any liquid or gaseous material which is a liquid while under pressure and having a flash point below 73 °F (22.8 °C) and having a boiling point below 100 °F (37.8 °C). (Class 1A flammable liquids.)

Materials which on account of their physical form or environmental conditions can form exlplosive mixtures with air and which are readily dispersed in air, such as dusts of combustible solids and mists of flammable or combustible liquid droplets.

DEPARTMENT OF TRANSPORTATION

49 CFR 173.300   Definitions.

(b) **Flammable compressed gas.** Any compressed gas as defined in paragraph (a) of this section shall be classed as "flammable gas" if any one of the following occurs:

(1) Either a mixture of 13 percent or less (by volume) with air forms a flammable mixture or the flammable range with air is wider than 12 percent regardless of the lower limit. These limits shall be determined at atmospheric temperature and pressure. The method of sampling and test procedure shall be acceptable to the Bureau of Explosives and approved by the Director, OHMT.

(2) Using the Bureau of Explosives' Flame Projection Apparatus, the flame projects more than 18 inches beyond the ignition source with valve opened fully, or, the flame flashes back and burns at the valve with any degree of valve opening.

(3) Using the Bureau of Explosives' Open Drum Apparatus, there is any significant propagation of flame away from the ignition source.

(4) Using the Bureau of Explosives' Closed Drum Apparatus, there is any explosion of the vapor-air mixture in the drum.

ENVIRONMENTAL PROTECTION AGENCY

40 CFR 261.21   Characteristic of ignitability.

(a) A solid waste exhibits the characteristic of ignitability if a representative sample of the waste has any of the following properties:

(3) It is an **ignitable compressed gas** as defined in 49 CFR 173.300 and as determined by the test methods described in that regulation or equivalent test methods approved by the Administrator under under Sections 260.20 and 260.21.

OCCUPATIONAL SAFETY AND HEALTH ADMINISTRATION

29 CFR 1910.1200   Hazard Communication.

(c) Definitions.

"Flammable" means a chemical that falls into one of the following categories:

(i) **"Aerosol, flammable"** means an aerosol that, when tested by the method described in 16 CFR 1500.45 [*Method for determining extremely flammable and flammable contents of self-pressurized containers*], yields a flame projection exceeding 18 inches at full valve opening, or a flashback (a flame extending back to the valve) at any degree of valve opening;

(ii) **"Gas, flammable"** means:

(A) A gas that, at ambient temperature and pressure, forms a flammable mixture with air at a concentration of thirteen (13) percent by volume or less; or

(B) A gas that, at ambient temperature and pressure, forms a range of flammable mixtures with air wider than twelve (12) percent by volume, regardless of the lower limit;

NATIONAL FIRE PROTECTION ASSOCIATION

NFPA 704   Identification of the Fire Hazards of Materials

Chapter 3 Flammability Hazards

4   Materials which will rapidly or completely vaporize at atmospheric pressure and normal ambient temperature or which are readily dispersed in air, and which will burn readily. This degree should include:

Gases;

Cryogenic materials;

Any liquid or gaseous material which is a liquid while under pressure and having a flash point below 73 °F (22.8 °C) and having a boiling point below 100 °F (37.8 °C). (Class 1A flammable liquids.)

Materials which on account of their physical form or environmental conditions can form explosive mixtures with air and which are readily dispersed in air, such as dusts of combustible solids and mists of flammable or combustible liquid droplets.

DEFINITIONS OF "FLAMMABLE GAS" AND CORRESPONDING TERMS

OCCUPATIONAL SAFETY AND HEALTH ADMINISTRATION

29 CFR 1910.1200   Hazard Communication.

(c) Definitions.

"Flammable" means a chemical that falls into one of the following categories:

(ii) **"Gas, flammable"** means:

(A) A gas that, at ambient temperature and pressure, forms a flammable mixture with air at a concentration of thirteen (13) percent by volume or less; or

(B) A gas that, at ambient temperature and pressure, forms a range of flammable mixtures with air wider than twelve (12) percent by volume, regardless of the lower limit;

DEPARTMENT OF TRANSPORTATION

49 CFR 173.300   Definitions.

(b) **Flammable compressed gas.** Any compressed gas as defined in paragraph (a) of this section shall be classed as "flammable gas" if any one of the following occurs:

(1) Either a mixture of 13 percent or less (by volume) with air forms a flammable mixture or the flammable range with air is wider than 12 percent regardless of the lower limit. These limits shall be determined at atmospheric temperature and pressure. The method of sampling and test procedure shall be acceptable to the Bureau of Explosives and approved by the Director, OHMT.

(2) Using the Bureau of Explosives' Flame Projection Apparatus, the flame projects more than 18 inches beyond the ignition source with valve opened fully, or, the flame flashes back and burns at the valve with any degree of valve opening.

(3) Using the Bureau of Explosives' Open Drum Apparatus, there is any significant propagation of flame away from the ignition source.

(4) Using the Bureau of Explosives' Closed Drum Apparatus, there is any explosion of the vapor-air mixture in the drum.

ENVIRONMENTAL PROTECTION AGENCY

40 CFR 261.21   Characteristic of ignitability.

(a) A solid waste exhibits the characteristic of ignitability if a representative sample of the waste has any of the following properties:

(3) It is an **ignitable compressed gas** as defined in 49 CFR 173.300 and as determined by the test methods described in that regulation or equivalent test methods approved by the Administrator under Sections 260.20 and 260.21.

NATIONAL FIRE PROTECTION ASSOCIATION

NFPA 704   Identification of the Fire Hazards of Materials

Chapter 3 Flammability Hazards

4   Materials which will rapidly or completely vaporize at atmospheric pressure and normal ambient temperature or which are readily dispersed in air, and which will burn readily. This degree should include:

Gases;

Cryogenic materials;

Any liquid or gaseous material which is a liquid while under pressure and having a flash point below 73 °F (22.8 °C) and having a boiling point below 100 °F (37.8 °C). (Class 1A flammable liquids.)

Materials which on account of their physical form or environmental conditions can form explosive mixtures with air and which are readily dispersed in air, such as dusts of combustible solids and mists of flammable or combustible liquid droplets.

DEFINITIONS OF "FLAMMABLE LIQUID"
AND CORRESPONDING TERMS

DEPARTMENT OF TRANSPORTATION

49 CFR 173.115   Flammable, combustible, and pyrophoric liquids; definitions.

(a) **Flammable liquid.**

  (1) For the purposes of this subchapter, a flammable liquid means any liquid having a flash point below 100°F (37.8°C), with the following exceptions:

   (i) Any liquid meeting one of the definitions specified in Section 173.300 [*Gases*];

   (ii) Any mixture having one component or more with a flash point of 100°F (37.8°C) or higher, that makes up at least 99 percent of the total volume of the mixture;

(b) (2) For the purposes of this subchapter:

   (i) An aqueous solution containing 24 percent or less alcohol by volume is considered to have a flash point of no less than 100°F (37.8°C) if the remainder of the solution is not subject to this subchapter, and

   (ii) An aqueous solution containing 24 percent or less alcohol by volume is not subject to the requirements of this subchapter if it contains no less than 50 percent water and no material (other than the alcohol) which is subject to this subchapter.

OCCUPATIONAL SAFETY AND HEALTH ADMINISTRATION

29 CFR 1910.1200   Hazard Communication.

(c) Definitions.

   "Flammable" means a chemical that falls into one of the following **categories:**

   (iii) **"Liquid, flammable"** means any liquid having a flashpoint below 100°F (37.8°C), except any mixture having components with flashpoints of 100°F (37.8°C) or higher, the total of which make up 99 percent or more of the total volume of the mixture.

NATIONAL FIRE PROTECTION ASSOCIATION

NFPA 704   Identification of the Fire Hazards of Materials

Chapter 3 Flammability Hazards

4   Materials which will rapidly or completely vaporize at atmospheric pressure and normal ambient temperature or which are readily dispersed in air, and which will burn readily. This degree should include:

Gases;

Cryogenic materials;

Any liquid or gaseous material which is a liquid while under pressure and having a flash point below 73°F (22.8°C) and having a boiling point below 100°F (37.8°C). (Class 1A flammable liquids.)

Materials which on account of their physical form or environmental conditions can form explosive mixtures with air and which are readily dispersed in air, such as dusts of combustible solids and mists of flammable or combustible liquid droplets.

3   Liquids and solids that can be ignited under almost all ambient temperature conditions. Materials in this degree produce hazardous atmospheres with air under almost all ambient temperatures or, though unaffected by ambient temperatures, are readily ignited under almost all conditions. This degree should include:

Liquids having a flash point below 73°F (22.8°C) and having a boiling point at or above 100°F (37.8°C) and those liquids having a flash point at or above 73°F (22.8°C) and below 100°F (37.8°C). (Class 1B and Class 1C flammable liquids);

Solid materials in the form of coarse dusts which may burn rapidly but which generally do not form explosive atmospheres with air;

Solid materials in a fibrous or shredded form which may burn rapidly and create flash fire hazards, such as cotton, sisal and hemp;

Materials which burn with extreme rapidity, ususally by reason of self-contained oxygen (e.g., dry nitrocellulose and many organic peroxides);

Materials which ignite spontaneously when exposed to air.

ENVIRONMENTAL PROTECTION AGENCY

40 CFR 261.21   Characteristic of **ignitability.**

(a) A solid waste exhibits the characteristic of ignitability if a representative sample of the waste has any of the following properties:

(1) It is a **liquid,** other than an aqueous solution containing less than 24 percent alcohol by volume and has flash point less than 60 °C (140 °F), as determined by a Pensky-Martens Closed Cup Tester, using the test method specified in ASTM Standard D-93-79 or D-93-80 (incorporated by reference, see Section 260.11), or a Setaflash Closed Cup Tester, using the test method specified in ASTM Standard D-3278-78 (incorporated by reference, see Section 260.11), or as determined by an equivalent test method approved by the Administrator under procedures set forth in Sections 260.20 and 260.21.

DEFINITIONS OF "FLAMMABLE SOLID (DOT)" AND CORRESPONDING TERMS

DEPARTMENT OF TRANSPORTATION

40 CFR 173.150 **Flammable solid; definition.**

For the purpose of this subchapter, "Flammable solid" is any solid material, other than one classed as an explosive, which, under conditions normally incident to transportation is liable to cause fires through friction, retained heat from manufacturing or processing, or which can be ignited readily and when ignited burns so vigorously and persistently as to create a serious transportation hazard. Included in this class are spontaneously combustible and water-reactive materials.

[*DOT water reactive materials are covered in the section titled "Water Reactive Material (Solid)" rather than in this section.*]

OCCUPATIONAL SAFETY AND HEALTH ADMINISTRATION

29 CFR 1910.1200 Hazard Communication.

(c) Definitions.

"Flammable" means a chemical that falls into one of the following categories:

(iv) **"Solid, flammable"** means a solid, other than a blasting agent or explosive as defined in section 1910.109 (a), that is liable to cause fire through friction, absorption of moisture, spontaneous chemical change, or retained heat from manufacturing or processing, or which can be ignited readily and when ignited burns so vigorously and persistently as to create a serious hazard. A chemical shall be considered to be a flammable solid if, when tested by the method described in 16 CFR 1500.44 [*Method for determining extremely flammable solids*], it ignites and burns with a self-sustained flame at a rate greater than one-tenth of an inch per second along its major axis.

ENVIRONMENTAL PROTECTION AGENCY

40 CFR 261.21 Characteristic of **ignitability**.

(a) A solid waste exhibits the characteristic of ignitability if a representative sample of the waste has any of the following properties:

(2) It is **not a liquid** and is capable, under standard temperature and pressure, of causing fire through friction, absorption of moisture or spontaneous chemical changes and, when ignited, burns so vigorously and persistently that it creates a hazard.

NATIONAL FIRE PROTECTION ASSOCIATION

NFPA 704   Identification of the Fire Hazards of Materials

Chapter 3 Flammability Hazards

4   Materials which will rapidly or competely vaporize at atmospheric pressure and normal ambient temperature or which are readily dispersed in air, and which will burn readily. This degree should include:

Gases;

Cryogenic materials;

Any liquid or gaseous material which is a liquid while under pressure and having a flash point below 73 °F (22.8 °C) and having a boiling point below 100 °F (37.8 °C). (Class 1A flammable liquids.)

Materials which on account of their physical form or environmental conditions can form explosive mixtures with air and which are readily dispersed in air, such as dusts of combustible solids and mists of flammable or combustible liquid droplets.

3   Liquids and solids that can be ignited under almost all ambient temperature conditions. Materials in this degree produce hazardous atmospheres with air under almost all ambient temperatures or, though unaffected by ambient temperatures, are readily ignited under almost all conditions. This degree should include:

Liquids having a flash point below 73 °F (22.8 °C) and having a boiling point at or above 100 °F (37.8 °C) and those liquids having a flash point at or above 73 °F (22.8 °C) and below 100 °F (37.8 °C). (Class 1B and Class 1C flammable liquids);

Solid materials in the form of coarse dusts which may burn rapidly but which generally do not form explosive atmospheres with air;

Solid materials in a fibrous or shredded form which may burn rapidly and create flash fire hazards, such as cotton, sisal and hemp;

Materials which burn with extreme rapidity, usually by reason of self-contained oxygen (e.g., dry nitrocellulose and many organic peroxides);

Materials which ignite spontaneously when exposed to air.

2   Materials that must be moderately heated or exposed to relatively high ambient temperatures before ignition can occur. Materials in this degree would not under normal conditions form hazardous atmospheres with air, but under high ambient temperatures or under moderate heating may release vapor in sufficient quantities to produce hazardous atmospheres with air. This degree should include:

Liquids having a flash point above 100 °F (37.8 °C), but not exceeding 200 °F (93.4 °C);

Solids and semisolids which readily give off flammable vapors.

OCCUPATIONAL SAFETY AND HEALTH ADMINISTRATION

29 CFR 1910.1200   Hazard Communication.

(c) Definitions.

"Flammable" means a chemical that falls into one of the following categories:

(iv) **"Solid, flammable"** means a solid, other than a blasting agent or explosive as defined in section 1910.109 (a), that is liable to cause fire through friction, absorption of moisture, spontaneous chemical change, or retained heat from manufacturing or processing, or which can be ignited readily and when ignited burns so vigorously and persistently as to create a serious hazard. A chemical shall be considered to be a flammable solid if, when tested by the method described in 16 CFR 1500.44 [*Method for determining extremely flammable solids*], it ignites and burns with a self-sustained flame at a rate greater than one-tenth of an inch per second along its major axis.

ENVIRONMENTAL PROTECTION AGENCY

40 CFR 261.21   Characteristic of **ignitability.**

(a) A solid waste exhibits the characteristic of ignitability if a representative sample of the waste has any of the following properties:

(2) It is **not a liquid** and is capable, under standard temperature and pressure, of causing fire through friction, absorption of moisture or spontaneous chemical changes and, when ignited, burns so vigorously and persistently that it creates a hazard.

DEPARTMENT OF TRANSPORTATION

49 CFR 173.150   **Flammable solid;** definition.

For the purpose of this subchapter, "Flammable solid" is any solid material, other than one classed as an explosive, which, under conditions normally incident to transportation is liable to cause fires through friction, retained heat from manufacturing or processing, or which can be ignited readily and when ignited burns so vigorously and persistently as to create a serious transportation hazard. Included in this class are spontaneously combustible and water-reactive materials.

DOT 49 CFR 171.8   Definitions and abbreviations.

"Spontaneously combustible material (solid)" means a solid substance (including sludges and pastes) which may undergo spontaneous heating or self-ignition under conditions normally incident to transportation or which may upon contact with the atmosphere undergo an increase in temperature and ignite.

NATIONAL FIRE PROTECTION ASSOCIATION

NFPA 704   Identification of the Fire Hazards of Materials

Chapter 3 Flammability Hazards

4   Materials which will rapidly or completely vaporize at atmospheric pressure and normal ambient temperature or which are readily dispersed in air, and which will burn readily. This degree should include:

Gases;

Cryogenic materials;

Any liquid or gaseous material which is a liquid while under pressure and having a flash point below 73 °F (22.8 °C) and having a boiling point below 100 °F (37.8 °C). (Class IA flammable liquids.)

Materials which on account of their physical form or environmental conditions can form explosive mixtures with air and which are readily dispersed in air, such as dusts of combustible solids and mists of flammable or combustible liquid droplets.

3   Liquids and solids that can be ignited under almost all ambient temperature conditions. Materials in this degree produce hazardous atmospheres with air under almost all ambient temperatures or, though unaffected by ambient temperatures, are readily ignited under almost all conditions. This degree should include:

Liquids having a flash point below 73 °F (22.8 °C) and having a boiling point at or above 100 °F (37.8 °C) and those liquids having a flash point at or above 73 °F (22.8 °C) and below 100 °F (37.8 °C). (Class IB and Class IC flammable liquids);

Solid materials in the form of coarse dusts which may burn rapidly but which generally do not form explosive atmospheres with air;

Solid materials in a fibrous or shredded form which may burn rapidly and create flash fire hazards, such as cotton, sisal and hemp;

Materials which burn with extreme rapidity, usually by reason of self-contained oxygen (e.g., dry nitrocellulose and many organic peroxides);

Materials which ignite spontaneously when exposed to air.

Continued . . .

2   Materials that must be moderately heated or exposed to relatively high ambient temperatures before ignition can occur. Materials in this degree would not under normal conditions form hazardous atmospheres with air, but under high ambient temperatures or under moderate heating may release vapor in sufficient quantities to produce hazardous atmospheres with air. This degree should include:

Liquids having a flash point above 100 °F (37.8 °C), but not exceeding 200 °F (93.4 °C);

Solids and semisolids which readily give off flammable vapors.

DEFINITIONS OF "HIGHLY TOXIC" AND CORRESPONDING TERMS

OCCUPATIONAL SAFETY AND HEALTH ADMINISTRATION

Appendix A to 29 CFR 1910.1200   Health Hazard Definitions (Mandatory)

3. **Highly Toxic:** A chemical falling within any of the following categories:

   (a) A chemical that has a median lethal dose (LD_{50}) of 50 milligrams or less per kilogram of body weight when administered orally to albino rats weighing between 200 and 300 grams each.

   (b) A chemical that has a median lethal dose (LD_{50}) of 200 milligrams or less per kilogram of body weight when administered by continuous contact for 24 hours (or less if death occurs within 24 hours) with the bare skin of albino rabbits weighing between two and three kilograms each.

   (c) A chemical that has a median lethal concentration (LC_{50}) in air of 200 parts per million by volume or less of gas or vapor, or 2 milligrams per liter or less of mist, fume, or dust, when administered by continuous inhalation for one hour (or less if death occurs within one hour) to albino rats weighing between 200 and 300 grams each.

DEPARTMENT OF TRANSPORTATION

49 CFR 173.326   **Poison A.**

   (a) For the purpose of Parts 170-189 of this subchapter, extremely dangerous poisons, Class A, are poisonous gases or liquids of such nature that a very small amount of the gas, or vapor of the liquid, mixed with air is dangerous to life. This class includes the following:

   (1) Bromacetone.
   (2) Cyanogen.
   (3) Cyanogen chloride containing less than 0.9 percent water.
   (4) Diphosgene.
   (5) Ethyldichlorarsine.
   (6) Hydrocyanic acid.
   (7) [Reserved]
   (8) Methyldichlorarsine.
   (9) [Reserved]
   (10) Nitrogen peroxide (tetroxide).
   (11) [Reserved]
   (12) Phosgene (diphosgene).
   (13) Nitrogen tetroxide-nitric oxide mixtures containing up to 33.2 percent weight nitric oxide.

49 CFR 173.343   **Poison B.**

(a) For the purposes of Parts 170-189 of this subchapter and except as otherwise provided in this part, Class B poisons are those substances, liquid or solid (including pastes and semisolids), other than Class A poisons or Irritating materials, which are known to be so toxic to man as to afford a hazard to health during transportation; or which, in the absence of adequate data on human toxicity, are presumed to be toxic to man because they fall within any one of the following categories when tested on laboratory animals:

(1) Oral toxicity.   Those which produce death within 48 hours in half or more than half of a group of 10 or more white laboratory rats weighing 200 to 300 grams at a single dose of 50 milligrams or less per kilogram of body weight, when administered orally.

(2) Toxicity on inhalation.   Those which produce death within 48 hours in half or more than half of a group of 10 or more white laboratory rats weighing 200 to 300 grams, when inhaled continuously for a period of one hour or less at a concentration of 2 milligrams or less per liter of vapor, mist, or dust, provided such concentration is likely to be encountered by man when the chemical product is used in any reasonable forseeable manner.

(3) Toxicity by skin absorption.   Those which produce death within 48 hours in half or more than half of a group of 10 or more rabbits tested at a dosage of 200 milligrams or less per kilogram body weight, when administered by continuous contact with the bare skin for 24 hours or less.

ENVIRONMENTAL PROTECTION AGENCY

40 CFR 261.11   Criteria for listing hazardous waste.

(a) The Administrator shall list a solid waste as a hazardous waste only upon determining that the solid waste meets one of the following criteria:

(2) It has been found to be fatal to humans in low doses or, in the absence of data on human toxicity, it has been shown in studies to have an oral LD 50 toxicity (rat) of less than 50 milligrams per kilogram, an inhalation LC 50 toxicity (rat) of less than 2 milligrams per liter, or a dermal LD 50 toxicity (rabbit) of less than 200 milligrams per kilogram or is otherwise capable of causing or significantly contributing to an increase in serious irreversible, or incapacitating reversible, illness. (Waste listed in accordance with these criteria will be designated **Acute Hazardous Waste.**)

NATIONAL FIRE PROTECTION ASSOCIATION

NFPA 704   Identification of the Fire Hazards of Materials

Chapter 2 Health Hazards

4   Materials which on very short exposure could cause death or major residual injury even though prompt medical treatment were given, including those which are too dangerous to be approached without specialized protective equipment. This degree should include:

Materials which can penetrate ordinary rubber protective clothing;

Materials which under normal conditions or under fire conditions give off gases which are extremely hazardous (i.e., toxic or corrosive) through inhalation or through contact with or absorption through the skin.

3   Materials which on short exposure could cause serious temporary or residual injury even though prompt medical treatment were given, including those requiring protection from all bodily contact. This degree should include:

Materials giving off highly toxic combustion products;

Materials corrosive to living tissue or toxic by skin absorption.

DEFINITIONS OF "IGNITABLE COMPRESSED GAS"
AND CORRESPONDING TERMS

ENVIRONMENTAL PROTECTION AGENCY

40 CFR 261.21   Characteristic of ignitability.

(a) A solid waste exhibits the characteristic of ignitability if a representative sample of the waste has any of the following properties:

(3) It is an **ignitable compressed gas** as defined in 49 CFR 173.300 and as determined by the test methods described in that regulation or equivalent test methods approved by the Administrator under Sections 260.20 and 260.21.

DEPARTMENT OF TRANSPORTATION

49 CFR 173.300   Definitions.

(a) **Compressed gas.**   The term "compressed gas" shall designate any material or mixture having in the container an absolute pressure exceeding 40 p.s.i. at 70°F or, regardless of the pressure at 70°F, having an absolute pressure exceeding 104 p.s.i. at 130°F; or any liquid flammable material having a vapor pressure exceeding 40 p.s.i. absolute at 100°F as determined by ASTM test D-323.

(b) **Flammable compressed gas.**   Any compressed gas as defined in paragraph (a) of this section shall be classed as "flammable gas" if any one of the following occurs:

(1) Either a mixture of 13 percent or less (by volume) with air forms a flammable mixture or the flammable range with air is wider than 12 percent regardless of the lower limit. These limits shall be determined at atmospheric temperature and pressure. The method of sampling and test procedure shall be acceptable to the Bureau of Explosives and approved by the Director, OHMT.

(2) Using the Bureau of Explosives' Flame Projection Apparatus, the flame projects more than 18 inches beyond the ignition source with valve opened fully, or, the flame flashes back and burns at the valve with any degree of valve opening.

(3) Using the Bureau of Explosives' Open Drum Apparatus, there is any significant propagation of flame away from the ignition source.

(4) Using the Bureau of Explosives' Closed Drum Apparatus, there is any explosion of the vapor-air mixture in the drum.

OCCUPATIONAL SAFETY AND HEALTH ADMINISTRATION

29 CFR 1910.1200   Hazard Communication.

(c) Definitions.

"Flammable" means a chemical that falls into one of the following categories:

(i) **"Aerosol, flammable"** means an aerosol that, when tested by the method described in 16 CFR 1500.45 [*Method for determining extremely flammable and flammable contents of self-pressurized containers*], yields a flame projection exceeding 18 inches at full valve opening, or a flashback (a flame extending back to the valve) at any degree of valve opening;

(ii) **"Gas, flammable"** means:

(A) A gas that, at ambient temperature and pressure, forms a flammable mixture with air at a concentration of thirteen (13) percent by volume or less; or

(B) A gas that, at ambient temperature and pressure, forms a range of flammable mixtures with air wider than twelve (12) percent by volume, regardless of the lower limit;

NATIONAL FIRE PROTECTION ASSOCIATION

NFPA 704   Identification of the Fire Hazards of Materials

Chapter 3 Flammability Hazards

4   Materials which will rapidly or completely vaporize at atmospheric pressure and normal ambient temperature or which are readily dispersed in air, and which will burn readily. This degree should include:

Gases;

Cryogenic materials;

Any liquid or gaseous material which is a liquid while under pressure and having a flash point below 73 °F (22.8 °C) and having a boiling point below 100 °F (37.8 °C). (Class IA flammable liquids.)

Materials which on account of their physical form or environmental conditions can form explosive mixtures with air and which are readily dispersed in air, such as dusts of combustible solids and mists of flammable or combustible liquid droplets.

ENVIRONMENTAL PROTECTION AGENCY

40 CFR 261.21   Characteristic of **ignitability.**

(a) A solid waste exhibits the characteristic of ignitability if a representative sample of the waste has any of the following properties:

(1) It is a **liquid,** other than an aqueous solution containing less than 24 percent alcohol by volume and has flash point less than 60 °C (140 °F), as determined by a Pensky-Martens Closed Cup Tester, using the test method specified in ASTM Standard D-93-79 or D-93-80 (incorporated by reference, see Scetion 260.11), or a Setaflash Closed Cup Tester, using the test method specified in ASTM Standard D-3278-78 (incorporated by reference, see Section 260.11), or as determined by an equivalent test method approved by the Administrator under procedures set forth in Sections 260.20 and 260.21.

DEPARTMENT OF TRANSPORTATION

49 CFR 173.115   Flammable, combustible, and pyrophoric liquids; definitions.

(a) **Flammable liquid.**

(1) For the purposes of this subchapter, a flammable liquid means any liquid having a flash point below 100 °F (37.8 °C), with the following exceptions:

(i) Any liquid meeting one of the definitions specified in Section 173.300 [*Gases*];

(ii) Any mixture having one component or more with a flash point of 100 °F (37.8 °C) or higher, that makes up at least 99 percent of the total volume of the mixture;

(b) **Combustible liquid.**

(1) For the purposes of this subchapter, a combustible liquid is defined as any liquid that does not meet the definition of any other classification specified in this subchapter and has a flash point at or above 100 °F (37.8 °C) and below 200 °F (93.3 °C) except any mixture having one component or more with a flash point at 200 °F (93.3 °C) or higher, that makes up at least 99 percent of the total volume of the mixture.

(2) For the purposes of this subchapter:

    (i) An aqueous solution containing 24 percent or less alcohol by volume is considered to have a flash point of no less than 100 °F (37.8 °C) if the remainder of the solution is not subject to this subchapter, and

    (ii) An aqueous solution containing 24 percent or less alcohol by volume is not subject to the requirements of this subchapter if it contains no less than 50 percent water and no material (other than the alcohol) which is subject to this subchapter.

[*While U.S. DOT's definitions for "Flammable Liquid" and "Combustible Liquid" do not have a flash point of 140° F as a criterion, the hazard classification system developed by the International Maritime Organization (IMO) does use a flash point of 140°F as the upper boundary for Class 3.3 Flammable Liquids. The class includes liquids with a flash point of 73°F or above up to and including 140°F. The IMO Class for a variety of materials is indicated in the Optional Hazardous Materials Table (49 CFR 172.102).*]

OCCUPATIONAL SAFETY AND HEALTH ADMINISTRATION

29 CFR 1910.1200   Hazard Communication.

    (c) Definitions.

        "Combustible liquid" means any liquid having a flashpoint at or above 100 °F (37.8 °C) but below 200 °F (93.3 °C), except any mixture having components with flashpoints of 200 °F (93.3 °C), or higher, the total volume of which make up 99 percent or more of the total volume of the mixture.

    "Flammable" means a chemical that falls into one of the following categories:

        (iii) **"Liquid, flammable"** means any liquid having a flashpoint below 100 °F (37.8 °C), except any mixture having components with flashpoints of 100 °F (37.8 °C) or higher, the total of which make up 99 percent or more of the total volume of the mixture.

NATIONAL FIRE PROTECTION ASSOCIATION

NFPA 704   Identification of the Fire Hazards of Materials

Chapter 3 Flammability Hazards

    4   Materials which will rapidly or completely vaporize at atmospheric pressure and normal ambient temperature or which are readily dispersed in air, and which will burn readily. This degree should include:

Gases;

Cryogenic materials;

Any liquid or gaseous material which is a liquid while under pressure and having a flash point below 73 °F (22.8 °C) and having a boiling point below 100 °F (37.8 °C). (Class IA flammable liquids.)

Materials which on account of their physical form or environmental conditions can form explosive mixtures with air and which are readily dispersed in air, such as dusts of combustible solids and mists of flammable or combustible liquid droplets.

3 Liquids and solids that can be ignited under almost all ambient temeprature conditions. Materials in this degree produce hazardous atmospheres with air under almost all ambient temperatures or, though unaffected by ambient temperatures, are readily ignited under almost all conditions. This degree should include:

Liquids having a flash point below 73 °F (22.8 °C) and having a boiling point at or above 100 °F (37.8 °C) and those liquids having a flash point at or above 73 °F (22.8 °C) and below 100 °F (37.8 °C). (Class IB and Class IC flammable liquids);

Solid materials in the form of coarse dusts which may burn rapidly but which generally do not form explosive atmospheres with air;

Solid materials in a fibrous or shredded form which may burn rapidly and create flash fire hazards, such as cotton, sisal and hemp;

Materials which burn with extreme rapidity, usually by reason of self-contained oxygen (e.g., dry nitrocellulose and many organic peroxides);

Materials which ignite spontaneously when exposed to air.

2 Materials that must be moderately heated or exposed to relatively high ambient temperatures before ignition can occur. Materials in this degree would not under normal conditions form hazardous atmospheres with air, but under high ambient temperatures or under moderate heating may release vapor in sufficient quantities to produce hazardous atmospheres with air. This degree should include:

Liquids having a flash point above 100 °F (37.8 °C), but not exceeding 200 °F (93.4 °C);

Solids and semisolids which readily give off flammable vapors.

DEFINITIONS OF "IGNITABLE SOLID" AND CORRESPONDING TERMS

ENVIRONMENTAL PROTECTION AGENCY

40 CFR 261.11   Characteristic of **ignitability**.

(a) A solid waste exhibits the characteristic of ignitability if a representative sample of the waste has any of the following properties:

(2) It is **not a liquid** and is capable, under standard temperature and pressure, of causing fire through friction, absorption of moisture or spontaneous chemical changes and, when ignited, burns so vigorously and persistently that it creates a hazard.

OCCUPATIONAL SAFETY AND HEALTH ADMINISTRATION

29 CFR 1910.1200   Hazard Communication.

(c) Definitions.

"Flammable" means a chemical that falls into one of the following categories:

(iv) **"Solid, flammable"** means a solid, other than a blasting agent or explosive as defined in section 1910.109 (a), that is liable to cause fire through friction, absorption of moisture, spontaneous chemical change, or retained heat from manufacturing or processing, or which can be ignited readily and when ignited burns so vigorously and persistently as to create a serious hazard. A chemical shall be considered to be a flammable solid if, when tested by the method described in 16 CFR 1500.44 [*Method for determining extremely flammable and flammable solids*], it ignites and burns with a self-sustained flame at a rate greater than one-tenth of an inch per second along its major axis.

DEPARTMENT OF TRANSPORTATION

49 CFR 173.150   **Flammable solid;** definition.

For the purpose of this subchapter, "Flammable solid" is any solid material, other than one classed as an explosive, which, under conditions normally incident to transportation is liable to cause fires through friction, retained heat from manufacturing or processing, or which can be ignited readily and when ignited burns so vigorously and persistently as to create a serious transportation hazard. Included in this class are spontaneously combustible and water-reactive materials.

49 CFR 171.8   Definitions and abbreviations.

"**Spontaneously combustible material (solid)**" means a solid substance (including sludges and pastes) which may undergo spontaneous heating or self-ignition under conditions normally incident to transportation or which may upon contact with the atmosphere undergo an increase in temperature and ignite.

NATIONAL FIRE PROTECTION ASSOCIATION

NFPA 704   Identification of the Fire Hazards of Materials

Chapter 3 Flammability Hazards

4   Materials which will rapidly or completely vaporize at atmospheric pressure and normal ambient temperature or which are readily dispersed in air, and which will burn readily. This degree should include:

> Gases;
>
> Cryogenic materials;
>
> Any liquid or gaseous material which is a liquid while under pressure and having a flash point below 73°F (22.8°C) and having a boiling point below 100°F (37.8°C). (Class IA flammable liquids.)
>
> Materials which on account of their physical form or environmental conditions can form explosive mixtures with air and which are readily dispersed in air, such as dusts of combustible solids and mists of flammable or combustible liquid droplets.

3   Liquids and solids that can be ignited under almost all ambient temperature conditions. Materials in this degree produce hazardous atmospheres with air under almost all ambient temperatures or, though unaffected by ambient temperatures, are readily ignited under almost all conditions. This degree should include:

> Liquids having a flash point below 73°F (22.8°C) and having a boiling point at or above 100°F (37.8°C) and those liquids having a flash point at or above 73°F (22.8°C) and below 100°F (37.8°C). (Class IB and Class IC flammable liquids);
>
> Solid materials in the form of coarse dusts which may burn rapidly but which generally do not form explosive atmospheres with air;
>
> Solid materials in a fibrous or shredded form which may burn rapidly and create flash fire hazards, such as cotton, sisal and hemp;
>
> Materials which burn with extreme rapidity, usually by reason of self-contained oxygen (e.g., dry nitrocellulose and many organic peroxides);
>
> Materials which ignite spontaneously when exposed to air.

Continued...

2   Materials that must be moderately heated or exposed to relatively high ambient temperatures before ignition can occur. Materials in this degree would not under normal conditions form hazardous atmospheres with air, but under high ambient temperatures or under moderate heating may release vapor in sufficient quantities to produce hazardous atmospheres with air. This degree should include:

Liquids having a flash point above 100 °F (37.8 °C), but not exceeding 200 °F (93.4 °C);

Solids and semisolids which readily give off flammable vapors.

DEFINITIONS OF "IRRITANT"
AND CORRESPONDING TERMS

OCCUPATIONAL SAFETY AND HEALTH ADMINISTRATION

Appendix A to 29 CFR 1910.1200   Health Hazard Definitions (Mandatory)

4. **Irritant:** A chemical, which is not corrosive, but which causes a reversible inflammatory effect on living tissue by chemical action at the site of contact. A chemical is a skin irritant if, when tested on the intact skin of albino rabbits by the methods of 16 CFR 1500.41 for four hours exposure or by other appropriate techniques, it results in an empirical score of five or more. A chemical is an eye irritant if so determined under the procedure listed in 16 CFR 1500.42 or other appropriate techniques.

DEFINITIONS OF "IRRITATING MATERIALS"
AND CORRESPONDING TERMS

DEPARTMENT OF TRANSPORTATION

40 CFR 173.381   **Irritating materials;** definition and general packaging requirements.

(a) For the purpose of Parts 170 through 189 of this subchapter, an irritating material is a liquid or solid substance which upon contact with fire or when exposed to air gives off dangerous or intensely irritating fumes, such as brombenzylcyanide, chloracetophenone, diphenylaminechlorarsine, and diphenylchlorarsine, but not including any poisonous material, Class A.

NATIONAL FIRE PROTECTION ASSOCIATION

NFPA 704   Identification of the Fire Hazards of Materials

Chapter 2 Health Hazards

2   Materials which on intense or continued exposure could cause temporary incapacitation or possible residual injury unless prompt medical treatment is give, including those requiring use of respiratory protective equipment with independent air supply. This degree should include:

Materials giving off toxic combustion products;

Materials giving off highly irritating combustion products;

Materials which either under normal conditions or under fire conditions give off toxic vapors lacking warning properties.

DEFINITIONS OF "ORGANIC PEROXIDE" AND CORRESPONDING TERMS

DEPARTMENT OF TRANSPORTATION

40 CFR 173.151a   Organic peroxide; definition.

(a) An organic compound containing the bivalent -O-O- structure and which may be considered a derivative of hydrogen peroxide where one or more of the hydrogen atoms have been replaced by organic radicals must be classed as an organic peroxide unless:

    (1) The material meets the definition of an explosive A or explosive B, as prescribed in Subpart C of this part, in which case it must be classed as an explosive,

    (2) The material is forbidden to be offered for transportation according to Section 172.101 [*Purpose and use of hazardous materials table*] or Section 173.21 [*Forbidden materials and packages*] of this subchapter,

    (3) It is determined that the predominant hazard of the material containing an organic peroxide is other than that of an organic peroxide, or

    (4) According to data on file with the RSPA, has been determined that the material does not present a hazard in transportation.

OCCUPATIONAL SAFETY AND HEALTH ADMINISTRATION

29 CFR 1910.1200   Hazard Communication.

(c) Definitions.

    "Organic peroxide" means an organic compound that contains the bivalent -0-0- structure and which may be considered to be a structural derivative of hydrogen peroxide where one or both of the hydrogen atoms has been replaced by an organic radical.

ENVIRONMENTAL PROTECTION AGENCY

40 CFR 261.23   Characteristic of reactivity. [*Reactive (Normally Unstable)*]

(a) A solid waste exhibits the characteristic of reactivity if a representative sample of the waste has *any* of the following properties:

    (1) It is normally unstable and readily undergoes violent change without detonating.

NATIONAL FIRE PROTECTION ASSOCIATION

NFPA 704   Identification of the Fire Hazards of Materials

Chapter 3 Flammability Hazards

3   Liquids and solids that can be ignited under almost all ambient temperature conditions. Materials in this degree produce hazardous atmospheres with air under almost all ambient temperatures or, though unaffected by ambient temperatures, are readily ignited under almost all conditions. This degree should include:

Liquids having a flash point below 73 °F (22.8 °C) and having a boiling point at or above 100 °F (37.8 °C) and those liquids having a flash point at or above 73 °F (22.8 °C) and below 100 °F (37.8 °C). (Class IB and Class IC flammable liquids);

Solid materials in the form of coarse dusts which may burn rapidly but which generally do not form explosive atmospheres with air;

Solid materials in a fibrous or shredded form which may burn rapidly and create flash fire hazards, such as cotton, sisal and hemp;

Materials which burn with extreme rapidity, usually by reason of self-contained oxygen (e.g., dry nitrocellulose and many organic peroxides);

Materials which ignite spontaneously when exposed to air.

Chapter 4   Reactivity (Instability) Hazards

2   Materials which in themselves are normally unstable and readily undergo violent chemical change but do not detonate. This degree should include materials which can undergo chemical change with rapid release of energy at normal temperatures and pressures or which can undergo violent chemical change at elevated temperatures and pressures. It should also include those materials which may react violently with water or which may form potentially explosive mixtures with water.

DEFINITIONS OF "OXIDIZER"
AND CORRESPONDING TERMS

OCCUPATIONAL SAFETY AND HEALTH ADMINISTRATION

29 CFR 1910.1200   Hazard Communication.

(c) Definitions.

"**Oxidizer**" means a chemical other than a blasting agent or explosive as defined in Section 1910.109(a), that initiates or promotes combustion in other materials, thereby causing fire either of itself or through the release of oxygen or other gases.

DEPARTMENT OF TRANSPORTATION

49 CFR 173.151   **Oxidizer**; definition.

An oxidizer for the purpose of this subchapter is a substance such as a chlorate, permanganate, inorganic peroxide, or a nitrate, that yields oxygen readily to stimulate the combustion of organic matter.

ENVIRONMENTAL PROTECTION AGENCY

40 CFR 261.21   Characteristic of ignitability.

(a) A solid waste exhibits the characteristic of ignitability if a representative sample of the waste has any of the following properties:

   (4) It is an **oxidizer** as defined in 49 CFR 173.151.

NATIONAL FIRE PROTECTION ASSOCIATION

NFPA 704   Identification of the Fire Hazards of Materials

Chapter 5 Special Hazards

   OX Materials which possess oxidizing properties shall be identified by the letters OX.

DEFINITIONS OF "POISON A"
AND CORRESPONDING TERMS

DEPARTMENT OF TRANSPORTATION

49 CFR 173.326   Poison A.

(a) For the purpose of Parts 170-189 of this subchapter, extremely dangerous poisons, Class A, are poisonous gases or liquids of such nature that a very small amount of the gas, or vapor of the liquid, mixed with air is dangerous to life. This class includes the following:

(1) Bromacetone.
(2) Cyanogen.
(3) Cyanogen chloride containing less than 0.9 percent water.
(4) Diphosgene.
(5) Ethyldichlorarsine.
(6) Hydrocyanic acid.
(7) [Reserved]
(8) Methyldichlorarsine.
(9) [Reserved]
(10) Nitrogen peroxide (tetroxide).
(11) [Reserved]
(12) Phosgene (diphosgene).
(13) Nitrogen tetroxide-nitric oxide mixtures containing up to 33.2 percent weight nitric oxide.

OCCUPATIONAL SAFETY AND HEALTH ADMINISTRATION

Appendix A to 29 CFR 1910.1200   Health Hazard Definitions (Mandatory)

3. **Highly Toxic:**   A chemical falling within any of the following categories:

(a) A chemical that has a median lethal dose (LD_{50}) of 50 milligrams or less per kilogram of body weight when administered orally to albino rats weighing between 200 and 300 grams each.

(b) A chemical that has a median lethal dose (LD_{50}) of 200 milligrams or less per kilogram of body weight when administered by continuous contact for 24 hours (or less if death occurs within 24 hours) with the bare skin of albino rabbits weighing between two and three kilograms each.

(c) A chemical that has a median lethal concentration (LC_{50}) in air of 200 parts per million by volume or less of gas or vapor, or 2 milligrams per liter or less of mist, fume, or dust, when administered by continuous inhalation for one hour (or less if death occurs within one hour) to albino rats weighing between 200 and 300 grams each.

ENVIRONMENTAL PROTECTION AGENCY

40 CFR 261.11   Criteria for listing hazardous waste.

(a) The Administrator shall list a solid waste as a hazardous waste only upon determining that the solid waste meets one of the following criteria:

(2) It has been found to be fatal to humans in low doses or, in the absence of data on human toxicity, it has been shown in studies to have an oral LD 50 toxicity (rat) of less than 50 milligrams per kilogram, an inhalation LC 50 toxicity (rat) of less than 2 milligrams per liter, or a dermal LD 50 toxicity (rabbit) of less than 200 milligrams per kilogram or is otherwise capable of causing or significantly contributing to an increase in serious irreversible, or incapacitating reversible, illness. (Waste listed in accordance with these criteria will be designated **Acute Hazardous Waste.**)

NATIONAL FIRE PROTECTION ASSOCIATION

NFPA 704   Identification of the Fire Hazards of Materials

Chapter 2 Health Hazards

4   Materials which on very short exposure could cause death or major residual injury even though prompt medical treatment were given, including those which are too dangerous to be approached without specialized protective equipment. This degree should include:

Materials which can penetrate ordinary rubber protective clothing;

Materials which under normal conditions or under fire conditions give off gases which are extremely hazardous (i.e., toxic or corrosive) through inhalation or through contact with or absorption through the skin.

DEPARTMENT OF TRANSPORTATION

40 CFR 173.343   **Poison B.**

(a) For the purposes of Parts 170-189 of this subchapter and except as otherwise provided in this part, Class B poisons are those substances, liquid or solid (including pastes and semisolids), other than Class A poisons or Irritating materials, which are known to be so toxic to man as to afford a hazard to health during transportation; or which, in the absence of adequate data on human toxicity, are presumed to be toxic to man because they fall within any one of the following categories when tested on laboratory animals:

  (1) Oral toxicity.   Those which produce death within 48 hours in half or more than half of a group of 10 or more white laboratory rats weighing 200 to 300 grams at a single dose of 50 milligrams or less per kilogram of body weight, when administered orally.

  (2) Toxicity on inhalation.   Those which produce death within 48 hours in half or more than half of a group of 10 or more white laboratory rats weighing 200 to 300 grams, when inhaled continuously for a period of one hour or less at a concentration of 2 milligrams or less per liter of vapor, mist, or dust, provided such concentration is likely to be encountered by man when the chemical product is used in any reasonable forseeable manner.

  (3) Toxicity by skin absorption.   Those which produce death within 48 hours in half or more than half of a group of 10 or more rabbits tested at a dosage of 200 milligrams or less per kilogram body weight, when administered by continuous contact with the bare skin for 24 hours or less.

OCCUPATIONAL SAFETY AND HEALTH ADMINISTRATION

Appendix A to 29 CFR 1910.1200   Health Hazard Definitions (Mandatory)

  3. **Highly Toxic:** A chemical falling within any of the following categories:

  (a) A chemical that has a median lethal dose (LD_{50}) of 50 milligrams or less per kilogram of body weight when administered orally to albino rates weighing between 200 and 300 grams each.

  (b) A chemical that has a median lethal dose (LD_{50}) of 200 milligrams or less per kilogram of body weight when administered by continuous contact for 24 hours (or less if death occurs within 24 hours) with the bare skin of albino rabbits weighing between two and three kilograms each.

(c) A chemical that has a median lethal concentration (LC_{50}) in air of 200 parts per million by volume or less of gas or vapor, or 2 milligrams per liter or less of mist, fume, or dust, when administered by continuous inhalation for one hour (or less if death occurs within one hour) to albino rates weighing between 200 and 300 grams each.

ENVIRONMENTAL PROTECTION AGENCY

40 CFR 261.11   Criteria for listing hazardous waste.

(a) The Administrator shall list a solid waste as a hazardous waste only upon determing that the solid waste meets one of the following criteria:

(2) It has been found to be fatal to humans in low doses or, in the absence of data on human toxicity, it has been shown in studies to have an oral LD 50 toxicity (rat) of less than 50 milligrams per kilogram, an inhalation LC 50 toxicity (rat) of less than 2 milligrams per liter, or a dermal LD 50 toxicity (rabbit) of less than 200 milligrams per kilogram or is otherwise capable of causing or significantly contributing to an increase in serious irreversible, or incapacitating reversible, illness. (Waste listed in accordance with these criteria will be designated **Acute Hazardous Waste.**)

NATIONAL FIRE PROTECTION ASSOCIATION

NFPA 704   Identification of the Fire Hazards of Materials

Chapter 2 Health Hazards

4   Materials which on very short exposure could cause death or major residual injury even though prompt medical treatment were given, including those which are too dangerous to be approached without specialized protective equipment. This degree should include:

Materials which can penetrate ordinary rubber protective clothing;

Materials which under normal conditions or under fire conditions give off gases which are extremely hazardous (i.e., toxic or corrosive) through inhalation or through contact with or absorption through the skin.

3   Materials which on short exposure could cause serious temporary or residual injury even though prompt medical treatment were given, including those requiring protection from all bodily contact. This degree should include:

Materials giving off highly toxic combustion products;

Materials corrosive to living tissue or toxic by skin absorption.

DEFINITIONS OF "PYROPHORIC" AND CORRESPONDING TERMS

OCCUPATIONAL SAFETY AND HEALTH ADMINISTRATION

29 CFR 1910.1200   Hazard Communication.

(c) Definitions.

"**Pyrophoric**" means a chemical that will ignite spontaneously in air at a temperature of 130 °F (54.4 °C) or below.

DEPARTMENT OF TRANSPORTATION

40 CFR 171.8   Definitions and abbreviations.

"**Spontaneously combustible material (solid)**" means a solid substance (including sludges and pastes) which may undergo spontaneous heating or self-ignition under conditions normally incident to transportation or which may upon contact with the atmosphere undergo an increase in temperature and ignite. [*Materials that meet the criteria for "Spontaneously Combustible Material (Solid)" are included in the DOT "Flammable Solid" hazard class.*]

49 CFR 173.115   Flammable, combustible, and pyrophoric liquids; definitions.

(c) **Pyrophoric liquids.**

(1) A pyrophoric liquid is any liquid that ignites spontaneously in dry or moist air at or below 130 °F (54.5 °C). [*Materials that meet the criteria for "Pyrophoric Liquid" are included in the DOT "Flammable Liquid" hazard class.*]

ENVIRONMENTAL PROTECTION AGENCY

40 CFR 261.21   Characteristic of **ignitability.**

(a) A solid waste exhibits the characteristic of ignitability if a representative sample of the waste has any of the following properties:

(1) It is a **liquid,** other than an aqueous solution containing less than 24 percent alcohol by volume and has flash point less than 60 °C (140 °F), as determined by a Pensky-Martens Closed Cup Tester, using the test method specified in ASTM Standard D-93-79 or D-93-80, or a Setaflash Closed Cup Tester, using the test method specified in ASTM Standard D-3278-78, or as determined by an equivalent test method approved by the Administrator under procedures set forth in Sections 260.20 and 260.21.

(2) **It is not a liquid** and is capable, under standard temperature and pressure, of causing fire through friction, absorption of moisture or spontaneous chemical changes and, when ignited, burns so vigorously and persistently that it creates a hazard.

NATIONAL FIRE PROTECTION ASSOCIATION

NFPA 704   Identification of the Fire Hazards of Materials

Chapter 3 Flammability Hazards

3   Liquids and solids that can be ignited under almost all ambient temperature conditions. Materials in this degree produce hazardous atmospheres with air under almost all ambient temperatures or, though unaffected by ambient temperatures, are readily ignited under almost all conditions. This degree should include:

Liquids having a flash point below 73°F (22.8°C) and having a boiling point at or above 100°F (37.8°C) and those liquids having a flash point at or above 73°F (22.8°C) and below 100°F (37.8°C). (Class IB and Class IC flammable liquids);

Solid materials in the form of coarse dusts which may burn rapidly but which generally do not form explosive atmospheres with air;

Solid materials in a fibrous or shredded form which may burn rapidly and create flash fire hazards, such as cotton, sisal and hemp;

Materials which burn with extreme rapidity, usually by reason of self-contained oxygen (e.g., dry nitrocellulose and many organic peroxides);

Materials which ignite spontaneously when exposed to air.

DEPARTMENT OF TRANSPORTATION

49 CFR 173.115   Flammable, combustible, and pyrophoric liquids; definitions.

(c) **Pyrophoric liquids.**

(1) A pyrophoric liquid is any liquid that ignites spontaneously in dry or moist air at or below 130°F (54.5°C). [*Materials that meet the criteria for "Pyrophoric Liquid" are included in the DOT "Flammable Liquid" hazard class.*]

OCCUPATIONAL SAFETY AND HEALTH ADMINISTRATION

29 CFR 1910.1200   Hazard Communication.

(c) Definitions.

"Pyrophoric" means a chemical that will ignite spontaneously in air at a temperature of 130°F (54.4°C) or below.

ENVIRONMENTAL PROTECTION AGENCY

40 CFR 261.21   Characteristic of **ignitability.**

(a) A solid waste exhibits the characteristic of ignitability if a representative sample of the waste has any of the following properties:

(1) It is a **liquid,** other than an aqueous solution containing less than 24 percent alcohol by volume and has flash point less than 60°C (140°F), as determined by a Pensky-Martens Closed Cup Tester, using the test method specified in ASTM Standard D-93-79 or D-93-80 (incorporated by reference, see Section 260.11), or a Setaflash Closed Cup Tester, using the test method specified in ASTM Standard D-3278-78 (incorporated by reference, see Section 260.11), or as determined by an equivalent test method approved by the Administrator under procedures set forth in Sections 260.20 and 260.21.

NATIONAL FIRE PROTECTION ASSOCIATION

NFPA 704   Identification of the Fire Hazards of Materials

Chapter 3 Flammability Hazards

3   Liquids and solids that can be ignited under almost all ambient temperature conditions. Materials in this degree produce hazardous atmospheres with air under almost all ambient temperatures or, though unaffected by ambient temperatures, are readily ignited under almost all conditions. This degree should include:

Liquids having a flash point below 73 °F (22.8 °C) and having a boiling point at or above 100 °F (37.8 °C) and those liquids having a flash point at or above 73 °F (22.8 °C) and below 100 °F (37.8 °C). (Class IB and Class IC flammable liquids);

Solid materials in the form of coarse dusts which may burn rapidly but which generally do not form explosive atmospheres with air;

Solid materials in a fibrous or shredded form which may burn rapidly and create flash fire hazards, such as cotton, sisal and hemp;

Materials which burn with extreme rapidity, usually by reason of self-contained oxygen (e.g., dry nitrocellulose and many organic peroxides);

Materials which ignite spontaneously when exposed to air.

DEFINITIONS OF "REACTIVE (CYANIDE AND SULFIDE WASTES)" AND CORRESPONDING TERMS

ENVIRONMENTAL PROTECTION AGENCY

40 CFR 261.23   Characteristic of reactivity. [*Reactive (Cyanide and Sulfide Wastes)*]

(a) A solid waste exhibits the characteristic of reactivity if a representative sample of the waste has *any* of the following properties:

(5) It is a cyanide or sulfide bearing waste which, when exposed to pH conditions between 2 and 12.5, can generate toxic gases, vapors or fumes in a quantity sufficient to present a danger to human health or the environment.

ENVIRONMENTAL PROTECTION AGENCY

40 CFR 261.23   Characteristic of reactivity. [*Reactive (Explosive)*]

(a) A solid waste exhibits the characteristic of reactivity if a representative sample of the waste has *any* of the following properties:

   (6) It is capable of detonation or explosive reaction if it is subjected to a strong initiating source or if heated under confinement.

   (7) It is readily capable of detonation or explosive decomposition or reaction at standard temperature and pressure.

   (8) It is a forbidden explosive as defined in 49 CFR 173.51, or a Class A Explosive as defined in 49 CFR 173.53 or a Class B Explosive as defined in 49 CFR 173.88.

NATIONAL FIRE PROTECTION ASSOCIATION

NFPA 704   Identification of the Fire Hazards of Materials

Chapter 4 Reactivity (Instability) Hazards

   4   Materials which in themselves are readily capable of detonation or of explosive decomposition or explosive reaction at normal temperatures and pressures. This degree should include materials which are sensitive to mechanical or localized thermal shock at normal temperatures and pressures.

   3   Materials which in themselves are capable of detonation or explosive decomposition or explosive reaction but which require a strong initiating source or which must be heated under confinement before initiation. This degree should include materials which are sensitive to thermal or mechanical shock at elevated temperatures and pressures or which react explosively with water without requiring heat or confinement.

DEPARTMENT OF TRANSPORTATION

49 CFR 173.53   Definition of Class A explosives.

[*This section defines nine numbered types of Class A explosives as well as thirteen other types of explosives. All of the types are high explosives, detonating explosives or other explosives of maximum hazard and include dynamite, lead azide, mercury fulminate, black powder, blasting caps, detonators and detonating primers.*]

49 CFR 173.88   Definition of Class B explosives.

(a) Explosives, Class B, are defined as those explosives which in general function by rapid combustion rather than detonation and include some explosive devices such as special fireworks, flash powders, some pyrotechnic signal devices and liquid or solid propellant explosives which include some smokeless powders. These explosives are further specifically described in paragraphs (b) to (g) of this section.

49 CFR 173.100   Definition of Class C explosives.

(a) Explosives, Class C, are defined as certain types of manufactured articles which contain Class A, or Class B explosives, or both, as components but in restricted quantities, and certain types of fireworks. These explosives are further specifically described in this section.

49 CFR 173.114a   Blasting agents.

(a) Definition of a blasting agent. A blasting agent is a material for blasting which has been tested in accordance with paragraph (b) of this section and found to be so insensitive that there is very little probability of accidental initiation to explosion or of transition from deflagration to detonation.

OCCUPATIONAL SAFETY AND HEALTH ADMINISTRATION

29 CFR 1910.1200   Hazard Communication.

(a) Definitions.

"**Explosive**" means a chemical that causes a sudden, almost instantaneous release of pressure, gas, and heat when subjected to sudden shock, pressure, or high temperature.

DEFINITIONS OF "REACTIVE (NORMALLY UNSTABLE)" AND CORRESPONDING TERMS

ENVIRONMENTAL PROTECTION AGENCY

40 CFR 261.23   Characteristic of reactivity. [*Reactive (Normally Unstable)*]

(a) A solid waste exhibits the characteristic of reactivity if a representative sample of the waste has *any* of the following properties:

(1) It is normally unstable and readily undergoes violent change without detonating.

OCCUPATIONAL SAFETY AND HEALTH ADMINISTRATION

29 CFR 1910.1200   Hazard Communication.

(c) Definitions.

"Unstable (reactive)" means a chemical which in the pure state, or as produced or transported, will vigorously polymerize, decompose, condense, or will become self-reactive under conditions of shock[,] pressure or temperature.

NATIONAL FIRE PROTECTION ASSOCIATION

NFPA 704   Identification of the Fire Hazards of Materials

Chapter 4 Reactivity (Instability) Hazards

2   Materials which in themselves are normally unstable and readily undergo violent chemical change but do not detonate. This degree should include materials which can undergo chemical change with rapid release of energy at normal temperatures and pressures or which can undergo violent chemical change at elevated temperatures and pressures. It should also include those materials which may react violently with water or which may form potentially explosive mixtures with water.

DEFINITIONS OF "REACTIVE (WATER REACTIVE)"
AND CORRESPONDING TERMS

ENVIRONMENTAL PROTECTION AGENCY

40 CFR 261.23   Characteristic of reactivity. [*Reactive (Water Reactive)*]

(a) A solid waste exhibits the characteristic of reactivity if a representative sample of the waste has *any* of the following properties:

(2) It reacts violently with water.

(3) It forms potentially explosive mixtures with water.

(4) When mixed with water, it generates toxic gases, vapors or fumes in a quantity sufficient to present a danger to human health or the environment.

NATIONAL FIRE PROTECTION ASSOCIATION

NFPA 704   Identification of the Fire Hazards of Materials

Chapter 4 Reactivity (Instability) Hazards

3   Materials which in themselves are capable of detonation or explosive decomposition or explosive reaction but which require a strong initiating source or which must be heated under confinement before initiation. This degree should include materials which are sensitive to thermal or mechanical shock at elevated temperatures and pressures or which react explosively with water without requiring heat or confinement.

2   Materials which in themselves are normally unstable and readily undergo violent chemical change but do not detonate. This degree should include materials which can undergo chemical change with rapid release of energy at normal temperatures and pressures or which can undergo violent chemical change at elevated temperatures and pressures. It should also include those materials which may react violently with water or which may form potentially explosive mixtures with water.

Chapter 5 Special Hazards

W̶   Materials which demonstrate unusual reactivity with water shall be identified with the letter W with a horizontal line through the center (W̶).

OCCUPATIONAL SAFETY AND HEALTH ADMINISTRATION

29 CFR 1910.1200   Hazard Communication.

(c) Definitions.

"**Water-reactive**" means a chemical that reacts with water to release a gas that is either flammable or presents a health hazard.

DEPARTMENT OF TRANSPORTATION

49 CFR 173.150   **Flammable solid;** definition.

For the purpose of this subchapter, "Flammable solid" is any solid material, other than one classed as an explosive, which, under conditions normally incident to transportation is liable to cause fires through friction, retained heat from manufacturing or processing, or which can be ignited readily and when ignited burns so vigorously and persistently as to create a serious transportation hazard. Included in this class are spontaneously combustible and water-reactive materials.

49 CFR 171.8   Definitions and abbreviations.

"**Water reactive material (solid)**" means any solid substance (including sludges and pastes) which, by interaction with water, is likely to become spontaneously flammable or to give off flammable or toxic gases in dangerous quantities.

DEFINITIONS OF "SENSITIZER" AND CORRESPONDING TERMS

OCCUPATIONAL SAFETY AND HEALTH ADMINISTRATION

Appendix A to 29 CFR 1910.1200   Health Hazard Definitions (Mandatory)

(5) **Sensitizer:** A chemical that causes a substantial proportion of exposed people or animals to develop an allergic reaction in normal tissue after repeated exposure to the chemical.

DEFINITIONS OF "SPONTANEOUSLY COMBUSTIBLE MATERIAL (SOLID)" AND CORRESPONDING TERMS

DEPARTMENT OF TRANSPORTATION

DOT 49 CFR 171.81   Definitions and abbreviations.

"Spontaneously combustible material (solid)" means a solid substance (including sludges and pastes) which may undergo spontaneous heating or self-ignition under conditions normally incident to transportation or which may upon contact with the atmosphere undergo an increase in temperature and ignite. [*Materials that meet the criteria for "Spontaneously Combustible Material (Solid)" are included in the DOT "Flammable Solid" hazard class.*]

ENVIRONMENTAL PROTECTION AGENCY

40 CFR 261.21   Characteristic of ignitability.

(a) A solid waste exhibits the characteristic of ignitability if a representative sample of the waste has any of the following properties:

(2) It is not a liquid and is capable, under standard temperature and pressure, of causing fire through friction, absorption of moisture or spontaneous chemical changes and, when ignited, burns so vigorously and persistently that it creates a hazard.

OCCUPATIONAL SAFETY AND HEALTH ADMINISTRATION

29 CFR 1910.1200   Hazard Communication.

(c) Definitions.

"Flammable" means a chemical that falls into one of the following categories:

(iv) **"Solid, flammable"** means a solid, other than a blasting agent or explosive as defined in section 1910.109 (a), that is liable to cause fire through friction, absorption of moisture, spontaneous chemical change, or retained heat from manufacturing or processing, or which can be ignited readily and when ignited burns so vigorously and persistently as to create a serious hazard. A chemical shall be considered to be a flammable solid if, when tested by the method described in 16 CFR 1500.44 [*Method for determining extremely flammable and flammable solids*], it ignites and burns with a self-sustained flame at a rate greater than one-tenth of an inch per second along its major axis.

NATIONAL FIRE PROTECTION ASSOCIATION

NFPA 704   Identification of the Fire Hazards of Materials

Chapter 3 Flammability Hazards

3   Liquids and solids that can be ignited under almost all ambient temperature conditions. Materials in this degree produce hazardous atmospheres with air under almost all ambient temperatures or, though unaffected by ambient temperatures, are readily ignited under almost all conditions. This degree should include:

Liquids having a flash point below 73 °F (22.8 °C) and having a boiling point at or above 100 °F (37.8 °C) and those liquids having a flash point at or above 73 °F (22.8 °C) and below 100 °F (37.8 °C). (Class IB and Class IC flammable liquids);

Solid materials in the form of coarse dusts which may burn rapidly but which generally do not form explosive atmospheres with air;

Solid materials in a fibrous or shredded form which may burn rapidly and create flash fire hazards, such as cotton, sisal and hemp;

Materials which burn with extreme rapidity, usually by reason of self-contained oxygen (e.g., dry nitrocellulose and many organic peroxides);

Materials which ignite spontaneously when exposed to air.

OCCUPATIONAL SAFETY AND HEALTH ADMINISTRATION

Appendix A to 29 CFR 1910.1200   Health Hazard Definitions (Mandatory)

6. **Toxic:** A chemical falling within any of the following categories:

   (a) A chemical that has a median lethal dose (LD_{50}) of more than 50 milligrams per kilogram but not more than 500 milligrams per kilogram of body weight when administered orally to albino rats weighing between 200 and 300 grams each.

   (b) A chemical that has a median lethal dose (LD_{50}) of more than 200 milligrams per kilogram but not more than 1,000 milligrams per kilogram of body weight when administered by continuous contact for 24 hours (or less if death occurs within 24 hours) with the bare skin of albino rabbits weighing between two and three kilograms each.

   (c) A chemical that has a median lethal concentration (LC_{50}) in air of more than 200 parts per million but not more than 2,000 parts per million by volume of gas or vapor, or more than 2 milligrams per liter but not more than 20 milligrams per liter of mist, fume, or dust, when administered by continuous inhalation for one hour (or less if death occurs within one hour) to albino rats weighing between 200 and 300 grams each.

DEFINITIONS OF "TOXIC WASTE"
AND CORRESPONDING TERMS

ENVIRONMENTAL PROTECTION AGENCY

40 CFR 261.11   Criteria for listing hazardous waste.

(a) The Administrator shall list a solid waste as a hazardous waste only upon determining that the solid waste meets one of the following criteria:

(3) It contains any of the toxic constituents listed in Appendix VIII....

Substances will be listed on Appendix VIII only if they have been shown in scientific studies to have toxic, carcinogenic, mutagenic or teratogenic effects on humans or other life forms. (Wastes listed in accordance with these criteria will be designated **Toxic Wastes**.)

DEFINITIONS OF "UNSTABLE (REACTIVE)" AND CORRESPONDING TERMS

OCCUPATIONAL SAFETY AND HEALTH ADMINISTRATION

29 CFR 1910.1200   Hazard Communication.

(c) Definitions.

"**Unstable (reactive)**" means a chemical which in the pure state, or as produced or transported, will vigorously polymerize, decompose, condense, or will become self-reactive under conditions of shock[,] pressure or temperature.

NATIONAL FIRE PROTECTION ASSOCIATION

NFPA 704   Identification of the Fire Hazards of Materials

Chapter 4 Reactivity (Instability) Hazards

Definitions.

> **Unstable** materials are those which in the pure state or as commercially produced will vigorously polymerize, decompose or condense or become self-reactive and undergo other violent chemical changes.

2   Materials which in themselves are normally unstable and readily undergo violent chemical change but do not detonate. This degree should include materials which can undergo chemical change with rapid release of energy at normal temperatures and pressures or which can undergo violent chemical change at elevated temperatures and pressures. It should also include those materials which may react violently with water or which may form potentially explosive mixtures with water.

ENVIRONMENTAL PROTECTION AGENCY

40 CFR 261.23   Characteristic of reactivity. [*Reactive (Normally Unstable)*]

(a) A solid waste exhibits the characteristic of reactivity if a representative sample of the waste has *any* of the following properties:

(1) It is normally unstable and readily undergoes violent change without detonating.

DEFINITIONS OF "WATER REACTIVE"
AND CORRESPONDING TERMS

OCCUPATIONAL SAFETY AND HEALTH ADMINISTRATION

29 CFR 1910.1200   Hazard Communication.

(c) Definitions.

"Water-reactive" means a chemical that reacts with water to release a gas that is either flammable or presents a health hazard.

DEPARTMENT OF TRANSPORTATION

49 CFR 173.150   **Flammable solid;** definition.

For the purpose of this subchapter, "Flammable solid" is any solid material, other than one classed as an explosive, which, under conditions normally incident to transportation is liable to cause fires through friction, retained heat from manufacturing or processing, or which can be ignited readily and when ignited burns so vigorously and persistently as to create a serious transportation hazard. Included in this class are spontaneously combustible and water-reactive materials.

49 CFR 171.8   Definitions and abbreviations.

"Water reactive material (solid)" means any solid substance (inlcuding sludges and pastes) which, by interaction with water, is likely to become spontaneously flammable or to give off flammable or toxic gases in dangerous quantities.

NATIONAL FIRE PROTECTION ASSOCIATION

NFPA 704   Identification of the Fire Hazards of Materials

Chapter 4 Reactivity (Instability) Hazards

3   Materials which in themselves are capable of detonation or explosive decomposition or explosive reaction but which require a strong initiating source or which must be heated under confinement before initiation. This degree should include materials which are sensitive to thermal or mechanical shock at elevated temperatures and pressures or which react explosively with water without requiring heat or confinement.

2 Materials which in themselves are normally unstable and readily undergo violent chemical change but do not detonate. This degree should include materials which can undergo chemical change with rapid release of energy at normal temperatures and pressures or which can undergo violent chemical change at elevated temperatures and pressures. It should also include those materials which may react violently with water or which may form potentially explosive mixtures with water.

Chapter 5 Special Hazards

W Materials which demonstrate unusual reactivity with water shall be identified with the letter W with a horizontal line through the center (W̶).

ENVIRONMENTAL PROTECTION AGENCY

40 CFR 261.23 Characteristic of reactivity. [*Reactive (Water Reactive)*]

(a) A solid waste exhibits the characteristic of reactivity if a representative sample of the waste has *any* of the following properties:

(2) It reacts violently with water.

(3) It forms potentially explosive mixtures with water.

(4) When mixed with water, it generates toxic gases, vapors or fumes in a quantity sufficient to present a danger to human health or the environment.

DEFINITIONS OF "WATER REACTIVE MATERIAL (SOLID)" AND CORRESPONDING TERMS

DEPARTMENT OF TRANSPORTATION

49 CFR 171.8   Definitions and abbreviations.

"**Water reactive material (solid)**" means any solid substance (including sludges and pastes) which, by interaction with water, is likely to become spontaneously flammable or to give off flammable or toxic gases in dangerous quantities. [*Materials that meet the criteria for "Water Reactive Material (Solid)" are included in the DOT "Flammable Solid" hazard class.*]

OCCUPATIONAL SAFETY AND HEALTH ADMINISTRATION

29 CFR 1910.1200   Hazard Communication.

(c) Definitions.

"**Water-reactive**" means a chemical that reacts with water to release a gas that is either flammable or presents a health hazard.

ENVIRONMENTAL PROTECTION AGENCY

40 CFR 261.23   Characteristic of reactivity. [**Reactive (Water Reactive)**]

(a) A solid waste exhibits the characteristic of reactivity if a representative sample of the waste has *any* of the following properties:

(2) It reacts violently with water.

(3) It forms potentially explosive mixtures with water.

(4) When mixed with water, it generates toxic gases, vapors or fumes in a quantity sufficient to present a danger to human health or the environment.

NATIONAL FIRE PROTECTION ASSOCIATION

NFPA 704   Identification of the Fire Hazards of Materials

Chapter 4 Reactivity (Instability) Hazards

3 Materials which in themselves are capable of detonation or explosive decomposition or explosive reaction but which require a strong initiating source or which must be heated under confinement before initiation. This degree should include materials which are sensitive to thermal or mechanical shock at elevated temperatures and pressures or which react explosively with water without requiring heat or confinement.

2 Materials which in themselves are normally unstable and readily undergo violent chemical change but do not detonate. This degree should include materials which can undergo chemical change with rapid release of energy at normal temperatures and pressures or which can undergo violent chemical change at elevated temperatures and pressures. It should also include those materials which may react violently with water or which may form potentially explosive mixtures with water.

Chapter 5 Special Hazards

W Materials which demonstrate unusual reactivity with water shall be identified with the letter W with a horizontal line through the center (W̶).

APPENDIX

DEPARTMENT OF TRANSPORTATION REGULATIONS ON
MULTIPLE LABELING
AND
THE HAZARD CLASS HIERARCHY

MULTIPLE LABELING

49 CFR 172.402   Additional labeling requirements.

(a) Multiple labeling.   Each package containing a material meeting the definition of more than one hazard class must be labeled as follows:

(1) A material classed as an Explosive A, Poison A, or Radioactive material that also meets the definition of another hazard class, must be labeled as required for each class.

(2) A Poison B liquid that also meets the definition of a Flammable liquid must be labeled "POISON" and "FLAMMABLE LIQUID."

(3) A material classed as Oxidizer, Flammable solid or Flammable liquid that also meets the definition of a Poison B must be labeled "POISON" in addituon to the class label.

(4) A material classed as a Flammable solid that also meets the definition of a water reactive material must have both the "FLAMMABLE SOLID" and "DANGEROUS WHEN WET" labels affixed.

(5) A material classed as a Corrosive material that also meets the definition of a Poison B shall be labeled with a "POISON" label in addition to the class label. This subparagraph does not apply to a material that would cause death due to corrosive destruction of tissue rather than by systematic poisoning.

(6) A material classed as a Posion B that also meets the definition of a corrosive material shall be labeled with a "CORROSIVE" label in addition to the class label.

(7) A material classed as a Flammable liquid that also meets the definition of a Corrosive material shall be labeled with a "CORROSIVE" label in addition to the class label.

(8) A material classed as a Flammable solid that also meets the definition of a Corrosive material shall be labeled with a "CORROSIVE" label in addition to the class label.

(9) A material classed as an Oxidizer that also meets the definition of a Corrosive material shall be labeled with a "CORROSIVE" label in addition to the class label.

(10) A material falling within the inhalation hazard criteria described in Section 173.3a(b)(2) shall be labeled with a "POISON" label in addition to any other label(s) required by this section. Duplication of the "POISON" label is not required.

(c) "DANGEROUS WHEN WET" label.   Each person who offers for transportation a package containing a hazardous material must affix to the package a "DANGEROUS WHEN WET" label as described in Section 172.423 when required by Section 172.101 [*Hazardous Materials Table*].

HAZARD CLASS HIERARCHY

49 CFR 173.2   Classification of a material having more than one hazard as defined in this part.

(a) Classification of material having more than one hazard as defined in this part. Except as provided in paragraph (b) of this section, a hazardous material, having more than one hazard as defined in this part, must be classed according to the following order of hazards:

(1) Radioactive material (except a limited quantity).

(2) Poison A.

(3) Flammable gas.

(4) Non-flammable gas.

(5) Flammable liquid.

(6) Oxidizer.

(7) Flammable solid.

(8) Corrosive material (liquid).

(9) Poison B.

(10) Corrosive material (solid).

(11) Irritating materials.

(12) Combustible liquid (in containers having capacities exceeding 110 gallons).

(13) ORM-B.

(14) ORM-A.

(15) Combustible liquid (in containers having capacities of 110 gallons or less).

(16) ORM-E.

(b) Exceptions.   Paragraph (a) of this section does not apply to —

(1) A material specifically identified in Section 172.101 [*Hazardous Materials Table*] of this subchapter;

(2) An explosive required to be classed and approved under Section 173.86 [*New Explosives Definitions; Approval and Notification*], or a blasting agent required to be classed and approved under Section 173.114a [*Blasting Agents*].

(3) An etiologic agent identified in Section 173.386 as those materials listed in 42 CFR 72.3; or

(4) An organic peroxide. (See Sections 172.101 and 173.151a of this subchapter.)

(5) A limited quantity radioactive material that also meets the definition of another hazard class (see Section 173.421-2).